First to Serve and Sacrifice

First to Serve and Sacrifice

Stories of Australians in the Boer War in Their Own Words

Daryl Moran

First Published in 2025 by Echo Books

Echo Books is an imprint of Superscript Publishing Pty Ltd,
ABN 76 644 812 395
PO Box 669, Woodend, Victoria, 3442
www.echobooks.com.au

Copyright © 2025 Daryl Moran

ISBN 978-1-923441-23-1

CONTENTS

FOREWORD

As one of the founding members of the Anglo Boer War Study Group, I visited South Africa in 1997 and 2000 and visited many battlefields where Australians fought between 1899–1902. As the author of *Second Harvest,* the compilation of the Boer War writings of Duncan Stock and his brother Tom, I particularly sought out Pink Hill near Colesberg, where Tom Stock was killed in action on 12 February 1900, becoming the first Caulfield Grammarian to die in war.

Having served as the Chair of the National Boer War Memorial Association from 2012 onwards, I oversaw the dedication and opening of the Memorial in Canberra in 2017. I also had the privilege of serving as the Chair of the sub-committee for the re-dedication of the South African Soldiers' Memorial in St Kilda Rd, Melbourne, in May 2025.

In his personal capacity and as a committee member of Military History and Heritage Victoria, Dr Daryl Moran has attended numerous Boer War services. On Boer War Day in 2024 at Melbourne's Shrine of Remembrance, he delivered the Commemoration Address about the *'Scottish Horse',* a mounted regiment in which both Caulfield and Malvern Grammarians served during the conflict.

His book *First to Serve and Sacrifice* tells in a chronological manner the life story, service and, in seven cases, the sacrifice of twenty-eight Caulfield and Malvern Grammarians during the Boer War. Extracts from their long-forgotten letters, related newspaper articles and other sources, tell their stories as witnesses to bravery, death, misery and every other experience in between at the 'sharp end' of the conflict. Many photographs are included in the book, and of special interest are the 'passport-type' pictures of the soldiers. In addition to telling of their times in the Boer War, the author has also written about their involvement at school and, in most cases, traced their later lives until their 'end of days.'

Of special interest to readers, the author has in some detail outlined the way the deaths of seven Caulfield Grammarians were marked in the field in South Africa, in their home communities in Australia and at their old school.

I commend Dr Moran for his detailed research and for adeptly bringing these fading memories to a new generation of readers. This volume provides a very fitting addition to the stories of our young countrymen, who were the first to serve and sacrifice in Australia's foreign war so long ago.

Peter Wilmot
President of Victorian Boer War Association
Custodian of the Queen's South Africa medal of Tom Stock

ACKNOWLEDGEMENTS

This book would not have been completed without the help and assistance of a range of people over a considerable period of years.

The late John Price was of invaluable assistance in the early 1990s when I first began undertaking research for this book. At the time, he was the pre-eminent researcher in Melbourne on all matters, 'Boer War' and a visit to his house was to step into an extensive library curated by a kind and generous writer. He had just published his volume *Southern Cross Scots,* which told the story of Lord Tullibardine's Scottish Horse, a unit largely raised in Melbourne, and which included several Caulfield and Malvern Grammarians. We stand on the shoulders of those who have gone before, and John was always very fulsome with his assistance and scholarly advice, for which I am most grateful.

Peter Wilmot, the President of the Boer War Association (Victoria) has generously written the Foreword to this book, and I am also most grateful for his careful editing of my work, especially the chapters concerned with the action in the South African theatre. He is also the author of an extensive retelling of the writings of the Stock brothers, Duncan and Tom, and has generously allowed me to draw on this resource in telling the story of Tom Stock, a Caulfield Grammarian. My thanks are extended to him for his interest and encouragement.

Dr Robin Droogleever has also been most generous with his editing of my work and has been especially helpful with the spelling of South African place names. His major contribution to this field is the authorship of what have become the standard works of history of all five Victorian Contingents to the Boer War, as well as that of the 1st New South Wales Mounted Rifles and other units. Dr Droogleever's extensive and scholarly work has provided a sound basis for the many narratives related in this book, and my sincere thanks are extended to him.

Caulfield Grammar's history of sport author, Dr Ian Wilkinson, has provided wise advice and support regarding the material relating to the history of Caulfield Grammar School. I have been fortunate to be able to draw upon his work as regards the history of sport, and I thank him for taking the time to also carefully edit those chapters.

Ms Judith Gibson, in her role as the Archivist at Caulfield Grammar School, has always provided access to materials for research purposes and has provided timely, informed and sage advice.

To the Principals of CGS from 1990 to the current times, the late Rev. Angas Holmes, Mr Stephen Newton AO, Rev. Andrew Syme and Mr Ashleigh Martin, my thanks are extended for allowing me to make use of the CGS Archives and to undertake this publication.

My sincere thanks are also extended to CGS's Executive Director of Community Engagement, Ms Sue Sonego, and Ms Linda Sprott, the Executive Director of the Caulfield Grammarians Association (CGA), who have actively supported, in practical terms, the publishing of the book.

Finally, as always, my thanks and appreciation are extended to my family for their patience and support, especially to my wife, Jenny, for her sage advice and being my best sounding board!

Dr Daryl Moran

INTRODUCTION

As a teenage student at Caulfield Grammar School in the mid-1960s, I often passed the large and ornate marble memorial plaque to Tom Stock, which hung in the foyer of the War Memorial Hall. Tom Stock had been killed in action in 1900 in the Second Anglo-Boer War (ABW) and was indeed the first Caulfield Grammarian to die on active service. As young teenage boys do, I gave the plaque scant regard until I returned to the Caulfield Campus as a member of the CGS Executive in the 1990s and had reason to pay it a little more attention. One of the areas that fell under my purview was that of administering the CGS Archives, and I had reason to explore its holdings as my office dealt with queries from CGS families about their Grammarian ancestors. A query from a Sydney-based researcher revealed that from the time of the ABW in 1899–1902, CGS shared a common ex-student, Stanley Spencer Reid, with Scotch College in Melbourne. One thing led to another, and my office started to research the nine other Caulfield Grammarians we knew had served in the conflict in South Africa; four of them, including Tom Stock and Stanley Reid, were noted on the school's Roll of Honour. At about this time, my research into Ivor Wilkinson, at just nineteen years of age, the youngest Caulfield Grammarian to be killed in South Africa, led me to the Melbourne home of famed Boer War researcher, John Price. He had just published *Southern Cross Scots,* the story of Tullibardine's Horse, the largely locally raised regiment in which Wilkinson had served. John was incredibly generous with his advice and followed up on numerous leads about other Grammarians for me. John had travelled to South Africa and walked many of the battlefields and spent many hours in research in a time well before the internet was invented. Over thirty years ago, his work was seminal in shaping our ideas about the ABW and the involvement of Grammarians in that conflict.

Years may have passed on, and life led me on many different pathways, but my abiding interest in the Caulfield Grammarians of the Boer War remained, and I picked up the threads of their stories again in recent times. By the 21st century, innumerable resources were now available via the internet and with access to the original CGS Enrolment Rolls and armed with the names of the 800 or so students who were enrolled at CGS between 1881 and 1901, I was now able to cross-reference all these resources and identify at least 28 Caulfield and Malvern Grammarians who had served in the ABW. This was a far cry from the 9 that had always been believed to be the definite number. In addition, three forgotten Grammarians were identified as having died during the conflict, and so their names could be added to the CGS Honour Roll. Once again, the internet revealed copies of letters and articles about these men, in many cases recounting their wartime experiences in their own words. Incredibly, numerous photographs of many of these soldiers came to light, and it was a joy to be able to see the faces behind the names.

It has been my honour and privilege to be able to write about these Grammarians from so long ago and to bring their stories and the story of the very early years of CGS to life once more. It is salutary to remember that seven of them lost their lives in the conflict and that another three received serious, life-changing wounds. Sadly, another four who served in the ABW were to die in World War 1.

It might be easy for us to dismiss the ABW as irrelevant to us in the 21st century, but these twenty-eight of our fellow Grammarians did answer the call of their country and were prepared to put their lives in danger. We can better understand who we are as a country (and school), if we have a better idea of where we have come from and what has gone before.

'Time Goes Forever,

But Memory Remains.'

LEST WE FORGET.

Dr Daryl Moran

AUTHOR'S NOTES

Given the passage of time and a lack of recorded information, there might be more Caulfield and Malvern Grammarians yet to be identified as having served in the ABW. I leave that task to future researchers and writers. My best efforts have been made to list details about the 28 soldiers identified to date, and these are listed in a summary Nominal Roll at the back of this book.

The years of attendance at CGS by these Grammarians are listed, but in some cases, their last year of attendance is an estimate, as the Enrolment Register does not record that detail. Similar details for Malvern Grammar are non-existent, and years of attendance are an estimate only.

Place names for locations in South Africa are as they appear in sources at the time.

For simplicity and consistency, I have retained the old measurement standards of the time, for example:

10 mile = 16 kilometres
10 feet = 3 metres
10 gallons = 38 litres
10 lbs (pounds) = 4.5 kilograms

GLOSSARY OF TERMS

Afrikaner   one descended from the settlers of Dutch, Huguenot, Flemish or early German heritage and living in South Africa

Afrikaans   the language spoken by the Afrikaners, which evolved from Dutch

Bivouac   a small temporary military camp

Boer   British term used to describe the Afrikaners as a 'farmer.'

Commando   Boer voluntary military units of guerilla militia

Kopje   a small hill

Kraal   African settlement or cattle enclosure

Laager   encampment

Picquet   a small group of soldiers performing sentry duty

'Uitlanders'   outlanders or ex-patriate immigrant outsiders

Veldt   the open countryside

MAP OF SOUTHERN AFRICA 1895–1902

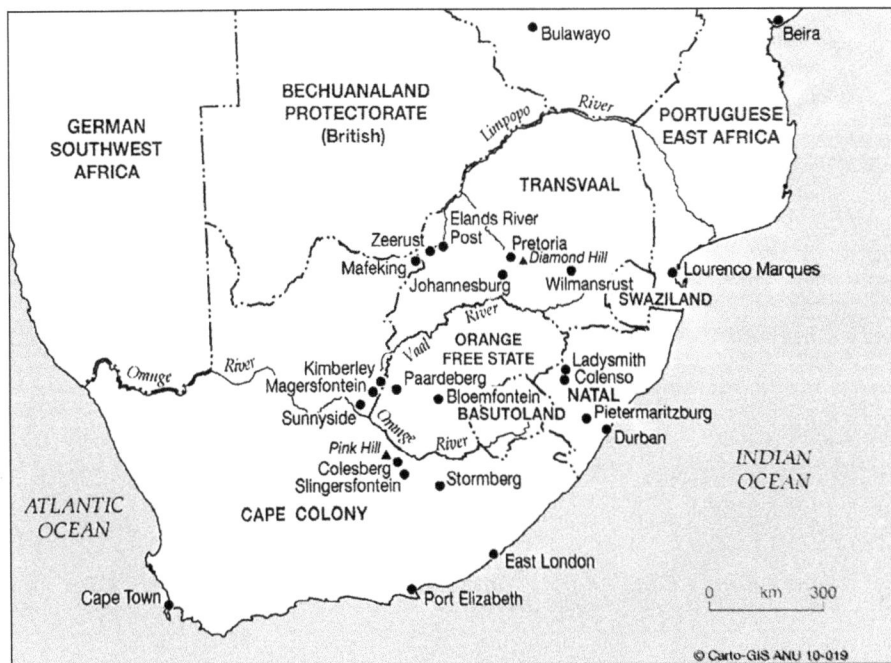

PREAMBLE – SETTING THE SCENE

1881

Sunday 27 February at Majuba Hill in Natal, South Africa.

The 400 Boer troops of Piet Joubert surrounded a British force, under Sir George Colley which was occupying Majuba Hill and caused 283 casualties out of 400 British troops, virtually annihilating the force.[1]
– John Phillips Kenyon, *Dictionary of British History.*

Saturday 26 March at 'Villa Marina' in St Kilda, Australia.

Have decided to open a school on my own account.[2]
– Joseph Henry Davies, founder of Caulfield Grammar School.

Within a month of each other in 1881, these two events had taken place in two different countries on two separate continents. The first event, in South Africa, involved the British Army suffering a catastrophic and humiliating defeat inflicted by Boer irregular troops, which in turn laid the grounds for a much larger conflict eighteen years later. The second event, in Australia a month later, marked the establishment of Caulfield Grammar School (CGS), now one of Australia's pre-eminent schools, and which, eighteen years later, saw twenty-four of its former students participate in that much larger South African conflict.

What historic circumstances in South Africa had led to the British Army suffering such a humiliating defeat? How did that defeat sow the seed for a much larger conflict eighteen years later. A brief background study of South Africa's history up 1899, helps to explain the historic circumstances which eventually led to the involvement of the young Australians.

In 1652 the Dutch East India Company established a shipping station at the Cape of Good Hope. It became populated with mainly Dutch and German settlers and farmers who depended on local African people as servants or slaves for their manual labourers. These European settlers called themselves 'Afrikaners' and they spoke a variation of the Dutch language called Afrikaans. In 1806, during the Napoleonic Wars, the British government took possession of the Cape Colony, mainly to protect its own sea lanes and world-wide commercial interests. The pre-eminent historian of the Boer War, Thomas Pakenham, provides a brief explanation of what followed.

Most of the Afrikaners, who remained the majority of the white population, were prepared to submit to British Crown rule, but a republican-minded minority, the Boers of the frontier, resented imperial interference, especially over their ill-treatment of their African [slaves] servants. In 1834 Britain ordered slaves to be emancipated in every part of the Empire. This precipitated the Great Trek: the exodus from the Cape in 1835–37 of about 5000 Boers and their African servants, across the Orange and Vaal rivers beyond the north-east frontiers of the Cape Colony.[3]

Boer commandoes in camp. (AWM P07379.079)

In 1843, the British established Natal, a second colony to the north-east of the Cape Colony. In the early 1850s the British government officially acknowledged the existence of the two Boer republics, the Transvaal and the Orange Free State. An uneasy situation continued to exist between the Boer and the British sides for another twenty years, until 1877 when the British annexed Transvaal as an initial step towards trying to form a united South Africa. In December 1880, under the leadership of its President Paul Kruger, the Boers in the Transvaal rose up in rebellion against the British. In what became known as the First Anglo-Boer War, the Boer forces inflicted a series of decisive defeats against the British which culminated in the Battle of Majuba Hill on 27 February 1881, and which led to the end of the First Anglo-Boer War. It was a crushing victory by the Boer forces over the British and highlighted many of the shortcomings of the

British Army which was still working on a model of warfare from the early 19th century. It had especially failed to grasp the implications of modern long-range rifles and associated guerilla tactics used by the skilled and usually horse-mounted Boer marksmen. Such was the humiliation of the defeat at Majuba Hill, that the British regiments involved were not allowed to claim it as a Battle Honour, as it represented one of the most humiliating defeats ever suffered by the British in their military history. For the Boer victors, it further served to reinforce their resentment of imperial interference and strengthen the belief in their righteous cause.

Although a small-scale encounter, it [Majuba Hill] persuaded the British to make

NOTABLE BOER WAR BRITISHERS

Arthur Conan Doyle was a British physician and writer who in 1887 created the Sherlock Holmes character. He served as a volunteer physician at the Langman Field Hospital at Blomfontein, between March and June 1900. He later published *The Great Boer War* a book arguing in favour of the British war aims. **Rudyard Kipling** was an English poet, journalist and novelist widely seen as the innovator of the short story genre. He immersed himself in the affairs of the Boer War, and visited many British hospitals, published supportive poetry and wrote numerous newspaper articles actively supporting the British cause. **Winston Churchill,** later a famous wartime British Prime Minister, worked as a journalist in South Africa for the [London] *Morning Post* newspaper in 1899. Two weeks after arrival he was captured but soon escaped Boer captivity, later serving

as a lieutenant with the South African Light Horse. **Robert Baden-Powell** was a British Army officer who oversaw the defence of the town of Mafeking during its 217-day siege by Boer forces from October 1899 to May 1900. His experiences during the siege working with messenger boy volunteers, led him to later found the worldwide (Boy) Scouts movement in 1907. His sister then set up the (Girl) Guides in 1910. **Mahatma Gandhi** was Indian born but was working as a lawyer in South Africa at the time of the Boer War. He founded the Natal Indian Ambulance Corps to support the British war effort. It attracted over 1100 trained and medically certified recruits. Among the places they served as stretcher bearers and medics was at Colenso, where Caulfield Grammarian **George Macartney** was very seriously wounded.

NOTABLE BOER WAR AUSTRALIANS

Andrew Barton Paterson, known as **'Banjo', Paterson,** was a solicitor from Sydney who wrote the poems 'Waltzing Matilda' and 'The Man from Snowy River.' From November 1899 to August 1900, he worked as a journalist in South Africa for the *Sydney Morning Herald* and *Melbourne Argus*. **Edith Dickenson** was a journalist who accompanied a contingent of Australian nurses to South Africa and reported for the *Adelaide Advertiser*. Her reports about concentration camps caused much controversy in Australia. Harry Harbord Morant known as **Breaker Morant** had a mysterious background in England before he emigrated to Australia. He first served with a South Australia contingent before joining the Bushveldt Carbineers where controversial circumstances led to his trial and execution by British troops on 27 February 1902.

Marianne Rawson was born at Avenel in Victoria, but she trained as a nurse in England. In March 1900 she was appointed as the Superintendent of the very first party of Victorian nurses to accompany soldiers of the Third Victorian Bushmen's Contingent to the Boer War. Sixty Australian nurses served in various contingents. Following a brave rescue of wounded comrades under fire on 1 September 1900, **John Bisdee,** an officer, serving with the Tasmanian Imperial Bushmen, became the first Australian born soldier to receive the Victoria Cross. **Harold Edward Elliott (Pompey)** was born in Charlton and served as a corporal with the 4th Victorian Imperial Bushmen. For gallantry in action, he was awarded the DCM and MiD. In WWI he served at Gallipoli and as a Brigadier-General he later led Australian troops in France.

peace and accept the Boer self-rule. Victory gave the Boers great confidence in their military skills, which was reflected in their aggressive stance at the start of the Second Anglo-Boer War eighteen years later.[4]

The resulting compromise peace between the British and the Boers sowed the seeds for further conflict. The Boer province of Transvaal was granted full self-government in all its internal affairs but was deemed to be a British Colony. Coupled with the humiliation that was visited upon the British at Majuba Hill, this key factor would remain a major cause of intense friction and aggravation between Britain and the Boers, as Thomas Pakenham and other historians noted.

On their part, many British soldiers [and settlers] felt deeply humiliated by the settlement. Moreover, it served to quicken, as soon emerged, the rising spirit of jingoism [on both sides]. Such was the Abyss of Blunders.[5]

In terms of duration and numbers engaged on both sides less than a thousand in all, Majuba Hill was little more than a skirmish, albeit a military object lesson in minor tactics and morale. Yet its effects were far-reaching. For the victors each individual Boer's yearning for unfettered personal independence, became transformed into an aggressive and unified Afrikaner nationalism all over South Africa. On the vanquished [British] side it inflicted a festering wound of bitterness and humiliation which could be healed only by revenge.[6]

The discovery of diamonds in 1870 at Kimberley on the border of the Cape Colony and the Orange Free State and the later finds of huge gold deposits in the Transvaal, added further 'fuel to the fire' to the tensions between the Boer and the British. Large numbers of fortune hunters [outsiders] from around the world arrived in the Boer republics and many of them were British subjects, including substantial numbers of Australians. Fearing that they would be swamped in their own country, the Boer Republics were not prepared to easily grant citizenship or voting rights to these outsiders or 'uitlanders.' This only heightened tensions further and caused inflammatory incidents on both sides. The Boer were pushed to their limits and fearing a British invasion, the Transvaal's President Paul Kruger, issued a war ultimatum to the British on 9 October 1899. Britain, keen to avenge Majuba Hill and to gain access to the mineral rich Boer territories, declared war the next day. Accordingly on 11 October 1899, to try to head off an impending British invasion, the numerically superior Boer forces attacked the British colony of Natal. So began the Second Anglo-Boer War between Britain [and its Empire] and the two Boer Republics of the Transvaal and the Orange Free State. Why would Caulfield and Malvern (MGS) Grammarians living in another country and on another continent, ever be involved in dealing with the long-lasting implications and outcomes of the Battle of Majuba, with its military outcomes in October 1899? To better understand the reasons, it is important to examine the history of the first eighteen years or so of CGS and the motivation of Joseph Henry Davies, the founding principal, in establishing the school. What influence did the three headmasters in the first eighteen years of the school have on helping to frame the attitudes of the students? What general expectations were placed upon young men at the time by society, that assisted the school to fulfill its educational mission? What were the educational and curriculum factors at CGS that helped to facilitate their willingness to enlist for military service in South Africa?

The 3rd Victorian Bushmen's Contingent (3VIB) including Caulfield Grammarian James Christie and Malvern Grammarian Arnold Davies, ride through the junction at St Kilda before embarkation to South Africa on the *Euryalus* on 10 March 1900. (AWM P11654.006)

Endnotes

1   Kenyon. J.P. *The Wordsworth Dictionary of British History.* Market House Books Ltd. 1981. p288

2   Davies. J.H. *Personal Diary.* Caulfield Grammar School Archives

3   Pakenham. T. *The Boer War.* Folio Society edition. London. 1999. pxxi

4   Grant. R.G. (ed). *1001 Battles that changed the course of History.* Harper Collins Books. 2011. Sydney. p663

5   Pakenham. T. *The Boer War.* ibid., p19

6   Barthrop. M. *The Anglo-Boer Wars. The British and the Afrikaners 1815–1902.* Cassell. London 1991. p43

PART ONE
BEGINNINGS

CHAPTER 1

CAULFIELD GRAMMAR SCHOOL BEGINS – PREPARING SOLDIERS

Preamble

Twenty-three Caulfield Grammarians and one staff member are known to have enlisted for service in the Boer War of 1899–1902, with seven losing their lives as a result. Four Malvern Grammarians are known to have enlisted and served in the conflict. There may be more from both schools who enlisted in South Africa but remain unidentified. All the entry dates at CGS for these students are documented here, but not their date of leaving the school as the enrolment register does not list this information. In a few cases, it is possible from other sources to calculate their last year at CGS, with others not so. They have been highlighted in **bold** for their first mention in this and the following chapter.

The founding Headmaster of Caulfield Grammar School (CGS) Joseph Henry Davies was the eldest son of English parents who had come to Melbourne from New Zealand in 1860 at the time of the Māori Wars. Young Henry's formal school education ceased at the age of eleven, when he was then employed in his father's legal firm. Unexpectedly, at the age of twelve he became the male head of his family of three brothers and two sisters, when his father died. His father's former legal partners encouraged Henry to study for the law and as a diligent and capable student at the age of fifteen, he passed his matriculation examination and began his law studies at the University of Melbourne. His deeply religious family were members of a sect of the Plymouth Brethren, which believed that the spreading of the Word of the Gospel was paramount in life, and so, perhaps not unexpectedly, in 1875 at the age of nineteen he gave away his university studies.

> His favourite sister Sarah had been recruited to the work of the Church Missionary Society in India. She found the mission seriously understaffed and appealed to her brother to join her. Her appeal, coinciding with his own inclinations was construed as a call from God and he left for India.[1]

He had been commissioned for this work at St Mary's Anglican Church in Caulfield by the incumbent vicar Rev. Hussey Burgh Macartney, under the guiding effect of three key prevailing social influences which influenced Davies' life. Firstly, Macartney encouraged the strong development in Davies of an

Evangelical Christian conviction and a passion to take the love of Christ to all the world. Secondly, Macartney encouraged Davies to hold a core value that embraced Ecumenism, a conviction that Christians could and should work together in unity across denominational lines. Finally, the concept of Globalism was embodied in the work of Davies whose vision included not only God's work in Australia, but throughout the world. These three factors were to heavily influence the work of Davies the Headmaster, the culture of the early days of CGS and in turn transmitted to the boys.[2]

Unfortunately, his work in India was not to last very long and in 1878 he returned to Australia for health and family reasons. Rev. Macartney had spoken prophetic words at Davies' commissioning:

We will expect that you will exercise no ordinary influence over the young men of Australia, and you may return to us some day from India, to bless thousands.[3]

Davies still wanted to serve on the mission field but aged twenty-five and as the families' eldest surviving son, found he had to provide for his younger siblings, especially his unmarried sister, Mary. Soon after his return from India, he established a tutorial service and coached young men for the matriculation examination to enter the University of Melbourne. He resumed his legal studies and was always keen to return to the mission field, but to do so realised he needed another source of income. After receiving positive encouragement, especially from some persuasive friends, notably Rev. Macartney, his diary entry for 26 March 1881 outlined his solution.

Have decided to open a school on my own account.[4]

Advertisements appeared in the Melbourne newspapers from 1 April 1881 announcing the new school, its location, and its advantages. Just two days before Caulfield Grammar School began classes for the first time, the same publication announced:

Caulfield Grammar School. Adjoining Elsternwick Station. Principal J. Henry Davies. B.A. Exhibitioner and Honourman in Classics, Scholar in Natural Sciences, Melbourne University. The school though easily accessible by rail, possesses all the advantages of a country situation and there is excellent accommodation for a limited number of boarders. Quarter commences April 25. Prospectuses on application.[5]

On 25 April 1881 Davies opened Caulfield Grammar School (CGS) in the then semi-rural location of Elsternwick with twelve boys as his first pupils, three of them matriculation candidates he was tutoring. The premises of the school were housed in a two-storey shop at the corner of Selwyn Street and Glenhuntly Road, a short distance from the Elsternwick railway station. The upstairs storey was converted to a boarding house with his sister Mary appointed as its first matron.

His evangelical and educational intentions were clear from the outset when he wrote in the first school's prospectus of 1881 that the school would be conducted on Christian principles and regular Biblical instruction of a strictly denominational character would be given in every class. He went further in the 1883 prospectus, when he proclaimed that the Bible lesson was to be made the first lesson of each day. While sparing no effort for the mental and physical training of the boys, the first aim of Davies as Headmaster was that the school should be a thoroughly Christian one.[6]

Davies saw the importance of these spiritual precepts and allied academic success, but he was also aware of the benefits conferred on character by the participation in many games and cultural pursuits.[7] In time, Davies provided a substantial playground for CGS as he, like other late 19th century educators, favoured the idea of boys participating in organized games.[8] Following the practice which had developed in English 'public' schools such as Eton, Harrow, Rugby and Winchester, most Australian secondary schools also encouraged their students to take part in organized games.[9] Many Australian educators accepted the arguments that had been developed in England to justify this rise in 'athleticism.' Put simply, it was argued that organized games not only helped produce 'fit' young men but also generated 'school spirit' and taught boys fair play, teamwork and how to win and lose graciously.[10] CGS historian Horrie Webber records that the first athletics sports of the school were held on 13 December 1882 in the adjacent Elsternwick paddock of Mr Short. These annual events were staged not only for the sporting benefit of the boys, but also to provide a grand social occasion for parents and friends of the school at which to gather and socialise. It also provided a marvellous public marketing opportunity over the coming years for CGS. In 1882 the athletics sports attracted a crowd of between 700 and 800 people to 'Stanmer Park'.

> The first athletics sports of the school were held at 'Stanmer Park' adjoining the school grounds. An attraction was added to the proceedings by the presence of the Band of the Blind Asylum. The attendance was large, for, beside several inhabitants of the neighbourhood, many visitors came by train. The events were well contested, and altogether an enjoyable afternoon was spent by those present.[11]

In his writings, Davies mentions sporting activities such as athletic sports, paper chases, and cricket matches. Australian Rules football was played with enthusiasm with the now traditional blue and white colours being worn. Few records remain of any of the organised games against other schools during this time, although football matches were being played in 1888 against Queen's College, St Kilda and Alma Road Grammar School.

Davies also wrote about other school pursuits associated with military cadets. An increasing militarism in Australian society, and consequently schools, had seen the establishment of army cadet units. These had arisen in the 1860s and had grown out of the practice of students undertaking various forms of military drill, mainly in the form of marching. Historian Craig Stockings states:

> Of particular importance, was the appointment of headmasters to prominent private schools in the colonies, men who believed earnestly in the moral aims of the British public school system, to produce boys fit to take leadership in a Christian state and Empire, which tacitly at least, made them very receptive to the concept of cadets.[12]

Davies agreed with these sentiments and maintained his support for the introduction of school cadets and the establishment of the Caulfield Grammar School Cadet Unit in 1885 made it the fifth oldest in a Victorian school.[13] Lt. Col. Frederick Sargood, the Victorian Minister of Defence, who lived nearby at his 23-acre estate 'Ripponlea' in East St Kilda, was a close friend of CGS and had his sons enrolled as pupils at the school.

Sargood stated that he would look to cadets in the future as furnishing a most important recruitment ground for the militia forces[14] and that his aim in founding school cadets was.

> To bind together in one patriotic brotherhood the youth of this country so that, should occasion arise, they may be able in after years to defend their country with the most telling effect.[15]

The CGS Cadet Unit was reported in 1888 to be participating in regular parades, drilling, participating in musketry training and preparing for the annual camp at Langwarrin. The school magazine of that year reported that in late September, the unit under the command of CGS staff member Lt. Conway Goulden, marched to the Elsternwick Rifle Butts, '… where they did a little shooting; but we regret to say, the scores are not worth recording.'

By the end of 1881, the school's enrolment had grown to 32 students, and it became clear to Davies that CGS had outgrown its founding premises at the

SIR FREDERICK THOMAS SARGOOD (1834–1903)

Frederick Sargood was born in London and was the son of a merchant. After being educated in England, he arrived in Melbourne with his parents and five sisters in 1850. He joined his father's wholesale soft-goods firm of Sargood, King and Co and eventually became its leading partner. He became a member of the Victorian Legislative Council in 1874, but when his wife died in 1880, he returned to England with his nine children. He re-married, returned to Melbourne in 1882 and then held the seat of South Yarra in the Legislative Council from 1882 to 1901. Sargood had been a member of the Victorian Volunteer Artillery since 1859 and was described as one of the best rifle shots in Victoria. He had helped form the St Kilda Rifle Club in the same year. In 1868 his large mansion 'Ripponlea' surrounded by 7 acres of land in

Elsternwick was completed. It was really a large self-sufficient farm with an elaborate underground watering system and its own electricity supply. In time the mansion grew to have 33 rooms and the fledgling CGS Cadet Unit made good use of the property's rifle range for their training drills. In 1883 he became Victoria's first Minister of Defence and worked hard to expand the Victorian Navy, local fortifications and supplies of armaments. The formation of school cadets in Victoria in 1884 was seen to be one of his greatest achievements. He became the President of the Melbourne Chamber of Commerce and a board member of the Commercial Bank. A supporter of Federation, he was elected to the first Australian Senate in 1901and served until his death in 1903.

corner of Selwyn Street and Glenhuntly Road. Drastic times called for drastic measures.

Davies raised £2000 to buy five acres in Cotton Street (now Regent Street) to build £700 worth of classrooms. The building included one large and two small classrooms and two small offices for teachers. The land nestled between properties owned by the Sargoods, ['Ripponlea'] Moores and Shorts ['Stanmer Park'] and Caulfield Grammar School students moved there in 1882. But his little school premises were still too small, so in 'spite of considering ordination and India' he decided in 1883 to spend £2225 on a 'new schoolhouse' which was built adjacent to the old. The Glenhuntly Road shop continued as the boarding house until the end of 1884 when the new schoolhouse was ready.[16]

Thirty-two new names were added to the enrolment register in 1884 and the addition of the new classrooms and expanded boarding house was a drawcard for these prospective students. Academic subject choices in those early days were varied.

> The earliest syllabuses included compulsory Bible study for every class, Writing, Arithmetic, English, Natural Science, Chemistry (with practical experiments for the middle and upper forms), Physiology for the upper forms and exams every three weeks. Music, Shorthand and Drawing were available in 1883 as extra subjects. Carpentry was offered from 1888 and Gym. class from 1890. Botany and Mapping were introduced in the 1890s.[17]

In 1884 the school's enrolment numbers had risen to fifty-four boys and among the new students were two who were later to enlist for the Boer War. **Andrew Percival Anderson** (1884–89?) and **James William Campbell** (1884–91?). Andrew Anderson was born in Prahran on 24th August 1874, the son of Andrew and Frances Anderson, of 35 Riversdale Road Hawthorn and he was aged ten when he was enrolled in First Form (Year 7). The 1885 Speech night program listed him as the Second Form prize winner for reading and dictation and he was mentioned again in 1889 when his recitation of the poem 'A Dutchman's Mistake' caused much laughter.

James Campbell (1884–1890) was aged 8 years old when he entered CGS on 6 October 1884. He was the son of James Campbell, the Victorian Postmaster General who was living at 'Himalaya' in Tennyson Street, St Kilda. Mr Campbell presided over the CGS Speech Nights in 1885 and in 1892. The 1887 Speech Night program listed James as winning the Second Form prize for Reading and in 1890 as being awarded the Form IV prize for History and Geography. He was a competitor in the CGS Athletics sports of 1888, taking part in the one hundred Yards Flat Race in the under fourteen age group and the two hundred Yards in the Under 15 level. A member of the CGS Cadet Unit, Campbell won Mr A. C. Goulden's Prize for Military Drill in 1890.

At the beginning of 1885 with enrolments standing at 65 students, **James William Christie** (1885–86) aged fifteen from Cheltenham enrolled at CGS and stayed for the following year, winning the Fifth Form prize for Writing and second prizes for Euclid and Algebra. However, after matriculating from CGS in 1886, he transferred to Melbourne Grammar School for the years of 1887–8.

> It was common practice in the late 19th century for boys to begin their schooling

at a suburban private school close to home, and then to transfer to one of the church-affiliated 'public' schools to complete their studies.[18]

William Boyd (1886–90) was born at Inglewood in 1871 and entered CGS as a boarder at the age of 13 in February 1886 as his mother was then living in Drummond Street, Ballarat. William was a good academic student, and he won the Third Form prize for Latin and French in his first year at the school. In 1888 he was shown in the Honour List as coming second in Greek and Geometry and third in French in Form V. In 1889 he was again amongst the Speech Night prize winners, winning the Fourth Form senior gymnastics prize as well as the Form V language and science prizes and being noted as second in class to the Dux of Form V. It was on the sporting field however, that he made his mark, and rated mentions in the then school magazine, *The Cricket* in 1888 for his football achievements.

William Boyd was particularly noted as an outstanding footballer and was listed among the best players and goal kickers in matches in 1888 against Brighton Grammar School,[19] Scotch College (St Kilda),[20] and Brighton Grammar School in 1889. In the match against St Kilda and Alma Road Grammar School on Friday 17 August 1888 it was reported that:

> Boyd was awarded a free kick and passed the ball on to Bowen who scored the third goal to Caulfield. The most notable players were Crosby, Rowe, Boyd, Dougall and Fetherstonhaugh.

Again, another report stated after the 5 September match against Queen's College that:

> After some play on the wing, Boyd got a free kick and registered a second goal for Caulfield. Courtney, Rowe, Crosby, T Dougall, Fetherstonhaugh and Boyd played well for the winners Caulfield who scored 4 goals 15 behinds to Queen's 3 goals 11 behind.

Boyd was also involved in sports administration and on 17 October 1888, the election of the Sports Committee took place with the following boys being elected: J. Dougall (Honorary Secretary), H. Courtney, T. Rowe, W. Boyd and J. Crosby (Honorary Treasurer). Boyd was listed in the November edition of *The Cricket* as participating at the athletics sports in the 300-yard flat race, the 150 yards hurdles, the four hundred yards flat race and the 880 yards flat race. In other cultural pursuits of William Boyd's school life, the magazine reported very drily:

Advertisements received too late for classification; Wanted – A new tune for Boyd's fiddle.

Academically he performed well enough to enrol in an Arts degree at Melbourne University in 1891.

Harry James Goodrich (sometimes spelt Goodricke) **Cattanach** (1886–89) was born in 1876 at Edenhope the second son of H.J.C. Cattanach a Commission Agent of 'Tynefield' in Elsternwick. Harry was one of two boys and seven girls in his family and his younger brother Adam was also a Caulfield Grammarian. Harry was aged 10½ when he entered CGS on 9th February 1886 and the CGS Speech Night program of that year shows that he won the Second Form prize for Bible and Latin. At the 1887 Speech Night he gave a recitation entitled 'Leap for Life' and won the Third form prizes for French and Arithmetic. In 1888 at the CGS Athletics Sports he ran in the Under 14 100 yards flat race, the two hundred Yards flat race, the two hundred yards Siamese race for Under 15s, as well as the Open 400 yards and 880 yards flat races. He also competed in 1889.

Stanley Spencer Reid (1886) was born on 12th July 1872 in Swan Hill and was the eldest son of six children of Sibyl and the Reverend John Reid who was the Presbyterian Minister at Mansfield. Stanley was aged thirteen when he enrolled as Caulfield Grammar School student no. 170 in February 1886, but he only remained at Caulfield until Easter of that year, when he transferred to Scotch College, where he completed his schooling in 1895.

George Raleigh Stewart (1886 -?) was born in Stawell in 1875 and at the age of eleven was enrolled at CGS in February 1886, the son of Samuel Stewart, a squatter. His mother lived at Grosvenor St, East St Kilda and in 1887 he was named in the CGS Speech Night program for his recitation on the night of *William Tell,* but his final year at CGS is unknown.

CGS football team from the late 1880s. (*Fields at Play*, p25)

By 1888, although CGS enrolments stood at 88 boys, Davies was increasingly yearning to return to the overseas mission field and to raise the necessary funds began to seek prospective buyers for his now thriving school. That year saw the enrolment of the last three 'Boer War' boys under his Headmastership. **Elmslie Fayrer Hewitt** (1888 –?) was enrolled as a boarder from *Malahide*, Fingal, Tasmania, but his year of leaving is unknown. **Walter Laishley Spier** (1888) was born in South Melbourne in 1874 and was enrolled at CGS in 1888 as a 13-year-old boarder, as his parents Walter and Charlotte lived in Hunter's Hill in Sydney. However, in January 1889, as a 14-year-old he left CGS and was enrolled at Sydney Grammar School (SGS) where his parent's address was listed as Alexander Street, Lane Cove. During his time at SGS he appears to have become quite involved with sport and in 1892 was named in the school's Rugby 1st XV.

Arthur George Thomas Williams (1888 –?) was born at Mologa near Boort in Victoria in 1875 and was aged eleven when he entered CGS with his older brother Edward Tudor Williams in February 1888. Their father was Dr Edward Johnson Williams of 3 Clarendon Terrace, Clarendon St, Melbourne.

After devoting himself for seven years to establishing the ethos and successful reputation of the young Caulfield Grammar School, Joseph Henry Davies eventually gave up head-mastering and travelled as the first Australian missionary to Korea. At a farewell lunch on 24 August 1888 just prior to leaving, he reminded his audience that he had always endeavoured to give the school a decidedly Christian tone and felt sure that those endeavours had not been wasted. He is credited with being one of the founders of the Presbyterian church in Korea, but tragically, just a few weeks after his arrival in Korea to begin his missionary work, Davies succumbed to smallpox and pneumonia and died on 7 April 1889.[22]

The founding work at CGS of Joseph Henry Davies saw a philosophy laid down that was fundamentally Christian in outlook, global in vision, and that worked across man-made divisions for the achievement of a higher and greater good. With an emphasis on 'Muscular Christianity' and sport, allied with CGS being an active supporter of the cadet movement, it could be contended that this view of Davies extended in a wider context to a loyalty and devotion to the 'God, Queen, Country' principle of the British Empire.

Davies had sold the School to an Anglican clergyman, the Rev. Ernest Judd Barnett who was its Headmaster from 1888 until 1896. Barnett, despite having no formal training as a teacher, had grown up in a family whose father was the proprietor of a private school in Tasmania. The CGS centenary history book records that:

Barnett was a man suited to the task. Of broad churchmanship and already deeply committed to the cause of missionary work, he would have earned the approval of Davies' clients. His own background of school-mastering allowed him to add the wider culture and deeper understanding of school organization and finance and professional school-mastering in general, which CGS needed to allow it to develop beyond the limits to which Davies had taken it.[23]

The school magazine of the time records details about the new Headmaster's intentions upon announcing that CGS was moving to a new and larger nearby site at 'Halstead' when it stated that:

> Lovers of sports, however, will be glad to know that there are more than five and a half acres of land out of which a good football ground and cricket pitch can be carved by sacrificing a part of the present garden, which our Headmaster is prepared to do.[24]

This new building was in fact the former home of Mr Crosby a long-time supporter of CGS. The property was located near Glen Eira Road in Elsternwick and almost adjacent to the current site of the school. In addition to occupying the existing former Crosby family buildings, Barnett arranged for the construction of a new school building complete with an imposing tower, which was occupied by CGS in March 1889.

> A large dining hall, master's rooms, eight bedrooms, bathrooms and a lavatory were added, together with four classrooms and an attached office.[25]

The occupation of the new buildings was short-lived however, when the entire structure was burnt to the ground on Sunday 27 April 1890, in an apparent arson attack by a former disgruntled non-teaching employee. Fortunately, no lives were lost or injuries sustained and undeterred, Barnett used the insurance money and completely rebuilt the school building, bigger and better than before. Constructed in brick, it was designed to hold twice the number of pupils and included a large lecture hall and eight new classrooms.

> The fine two storey building with its square tower, that symbol of solidity and respect in Victorian times, became a landmark to be seen by passengers on the train between Balaclava and Elsternwick.[26]

The early 1890s was a time of economic turmoil in Victoria, especially when the Land Boom bust. Bankruptcies in Melbourne soared, several banks closed their doors, and the share market suffered a dramatic fall. Headmaster Barnett stood

MAJOR BUILDING FIRE AT CGS IN 1890

'The north-east wing of the Caulfield Grammar School on Glen Eira Road was destroyed by fire last Sunday night and the remainder of the building had a similar escape from the same fate. The school is a prominent landmark in the locality and stands in the centre of very large grounds. The Principal Mr Barnett was away in the city and Mr Lynch an Assistant Master was at St Mary's church with about 20 boarders. Miss Ellen McDonald, the Matron, one boy ill in bed with influenza and a few other boys were the only occupants, when flames were noticed coming from the Carpentry shop. Nothing could be done to stop the fire as there were no fire extinguishing appliances on the premises. While someone ran to alert the fire brigade as well as those at the church, it was seen that the large flames were well alight in the schoolroom, the dining room, the servant's room and the laundry. All the boys and few staff were joined by passersby in working like Trojans to go into the building and throw out through the windows the contents of the rooms. Most of the school's books and papers were saved, but sadly most of the boys lost all their clothing and possessions. After the fire the boarders were housed at the nearby 'Ripponlea' residence, the commodious home of Colonel Sargood. It is suspected that the fire was deliberately lit, as just before the fire a man was seen lurking about the premises. *Caulfield and Elsternwick Leader.* Saturday 3 May 1890. p5

firm in these troubled times and in 1893 he noted that numbers on the roll had been well maintained. CGS historian Horrie Webber observed:

To conduct a private school during these difficult times demanded faith, courage and perseverance; faith not only in the validity and viability of the school, but in the future of the [Victorian] colony.[27]

CGS enrolments dropped slightly to remain steady at the mid-seventies, a sign that Barnett's leadership of the school had staunch support in these difficult times. The stream of enrolments remained constant and in 1889 a total of twenty-eight new boys entered the school, including three sets of brothers. This meant the school now numbered ninety-two students, the largest number to date in CGS history. Two 'Boer War' students also enrolled in the next two years.

'Halstead' early 1890s. Note brick structure built after the fire in 1890.
(CGS Archives)

Andrew Percival Rowan (1889–90?) was born at St Kilda on 31st of March 1876 and was one of the seven sons and three daughters of Andrew and Margaret Rowan of Brighton Road, Elsternwick. He entered the school in 1889 as a thirteen-year-old day scholar and remained for at least two years. The 1889 CGS Speech Night program shows that Percy Rowan won the Form II prizes for Bible, Reading and Recitation, Grammar, and Geography prizes. He continued that good academic progress the following year when he was awarded the Form III Conduct prize. He was a member of the cadet unit and in 1890 he won the Trophy for Rifle Shooting, ironically a school prize donated by his mother.

The following year saw the enrolment of **Thomas Barham Foster** (1890–1891) who was born in Sale on 27th July 1875. As the son of William and Catherine Foster of 'Highwood' Fulton Street, East St Kilda he was enrolled at CGS in Form IVB aged 15 in 1890. In 1891, he gained third place in the Form V prize for Arts. The following year, his parents had relocated to Neill Street, Soldiers Hill, Ballarat and on 23 May 1892 he was enrolled at Ballarat College as student no. 798. Also enrolled in 1890 was **Charles Hugh Thomas** (1890–1894) who was born on 2nd March 1878 and was the son of Mary Sarah and John Hampden Thomas an accountant who lived in Orrong Road Caulfield. Charles was aged twelve when he entered CGS in February 1890. In his first year at the school, he was named as the Dux of the Third Form and in 1894 won the Form IVB first prize for science, the Form Conduct prize and came second in the Form in Mathematics.

CGS Boarders 1889. (CGS Archives)

In 1891 Principal Barnett had ensured that his school joined the amateur based Schools' Association of Victoria (SAV) for the purpose of engaging in organized cricket, football, and athletics competitions against local boys' schools in Melbourne and surrounds. In this instance, it was thought advantageous to have inter-school games and sporting meetings organized by a single body under an agreed set of rules and conditions. The founding schools were Caulfield Grammar School, Carlton College, Cumloden, Geelong College, Armadale High School, King's College, Kew High School, South Melbourne College, Camberwell Grammar School, Toorak College (then a boy's school) and Queen's College. It should be noted that the schools designated as High Schools were private schools and not Victorian Education Department secondary schools, as these did not come into existence until the early 1900s.

That following year saw the admission of **Bernard Everett Bardwell,** (1892–1896) the eldest son of Everett and Fanny Bardwell. Bernard was born on 8th September 1881 at Emerald Hill in Melbourne and was first educated privately, then at East Malvern Grammar School, before being enrolled at CGS in 1892 aged eleven. At Speech Night that year, he gained second place in both the Form Two Arts prize and the Mathematic prize. He left at the end of 1896 as his father, a lawyer, assumed a position at Mosman in Perth, West Australia in 1897.

The year of 1893 brought the enrolment of three boys who were to become

'Boer War' soldiers. **Edward (Ted) Arthur Duncan** (1893–1894) was born in Wandiligong, (Vic) in 1876, the middle son of Edward and Margaret Duncan and aged seventeen enrolled at CGS in 1893 and appears to have left at the end of 1894.

Thomas (Tom) Stock (1893–1896) was born on 12 August 1876 at Carapook near Casterton, Victoria and was the son of John and Christina Stock, farmers from Sandford. Tom was just over 16 years of age when he enrolled in 1893 as student number 411 and as a boarder. Tom showed aptitude and abilities in non-academic areas of school life and the CGS Speech Night program of 1894 showed that he won, *Mrs Were's Prize for Unselfish and Courteous Behaviour.* He excelled on the sporting field and was noted as an all-round athlete, and named as one of the best players as a member of the 1st XVIII, played in the 1893 match against Kew High School. The Report of the CGS Annual Athletics Sports held on 26th October 1894, showed that Tom competed in several events and gained first place in the 150 yards hurdles and the Master's Plate over 440 yards. He gained second places in kicking the football, the Open 100 yards and the Open High Jump and won enough aggregate points to be judged the second most successful athlete on the day. The following year, 1895, his last full year at CGS, he was again runner-up at the CGS Athletics Sports for the Senior Cup with victories in the Open 100 yards sprint, the one hundred yards Siamese race and second places in the Long Jump and Open Vaulting. At the same event he won the Kicking the Football competition with a distance of sixty-five yards or nearly 59 metres. Academic study, however, was not his strong point with his school record showing low positions and intermittent attendance in some years, and when he failed his Matriculation Certificate in May 1896, Tom left CGS to assume station life as a grazier at Sandford.

Frank Valentine Weir (1893) was born on 14 February 1878 at Deniliquin, NSW. He was the son of Joseph and Mrs Weir of *Ashton* Canterbury Road, Warrnambool and attended Melbourne Grammar School in his primary years and then entered CGS (no. 400) in 1893 aged fifteen. The following year, 1894, he returned home to complete four years of secondary schooling at Warrnambool College.

The year of 1895 brought the enrolments of **Reginald James John Holloway** (1895–1896) who was the son of George and Jane Holloway and was born at Duck Swamp near Minyip in Victoria on 13th of August 1880 and attended CGS as a boarder for two years.

Gerald Massey Ivor Wilkinson (1895–1896) was born on 12th May 1883 and was the youngest of eight children (four boys, four girls) and was named after the English poet Gerald Massey. Ivor or 'Jinks' as he was known, was the son of Robert and Anderina Wilkinson who lived first in Talbot and then in Maryborough (Vic). Ivor had attended primary school in Maryborough, as his father Robert conducted a pharmacy business there until his death in 1890, when Ivor was only seven years old.[29] Ivor was aged 12 when he enrolled at CGS in 1895 as student (496) and although the form results of that year show that his marks weren't high due to not attending school for some of the academic year, the 1896 CGS Speech Night program notes that he was awarded second place in the Form VB prize for Science. In later years he was remembered in articles in the school magazine by some former classmates.

> My memory is distinct of the quarter of an hour before school started that morning in February 1896, when we made friends with each other through the medium of handball. The figure I remember most in that short game was of a short stumpy fellow with freckled face and a close-cropped ginger hair – that was 'Jinks' Wilkinson – and he now sleeps soundly beneath the South African veldt – not the only CGS boy who rests there.[30] Jinks' Wilkinson – as fat as he was long, was another who made our lives merry and gay in the old school.[31]

A former school friend from his days at Maryborough Primary School, Reginald Greenwood, later remembered 'Jinks' in a local newspaper article in 1950.

> I was privileged to know 'Jinks' well in my early days at Maryborough and I often kicked the football with him in the back street behind High Street. He was then living with his mother and brothers in a new home which was erected across the road from the back gate of my own home. 'Jinks' at that time was attending a school (CGS) in Melbourne and came home on holidays at intervals. I well remember him as a well-built rugged type of lad who excelled in sport. I fancy he was inclined to be reddish in the thatch.[32]

Barnett also encouraged participation at CGS in carpentry, tennis, gymnastics, music, self-expression activities (acting), public speaking and the maintenance of high academic standards. Importantly he supported the continuation of the school's first cadet unit and actively promoted and encouraged its activities. A later observer noted that Victorians of the time were much concerned with military preparedness and CGS willingly played its part.[33] Spiritual concerns were also to the forefront of Barnett's philosophy and his Speech Night report of 1892 stated that he clearly saw the great practical benefit that accrued from the daily,

CGS SPORT 1881–1901

The role of sport at Caulfield Grammar School in the first twenty years of its existence was one of increasing importance. From its very early days under the Headship of J H Davies, the school had tried to make arrangements for recreation and organised games for the boys, by having a large school playground. Quite often, games were organised between various schools' boarding houses and it is possible that CGS was participating in these games as early as 1882. Annual CGS athletics sports days were held and became popular social events. Cricket, tennis and Australian football appeared to be the main organised sport for CGS boys in those first twenty years of its existence. For reasons of ease of travel and convenience, many of the early matches were played against local schools, many of which no longer exist. Nearby still extant schools Brighton Grammar School (BGS) and Haileybury College (HC) alone remain out of the many that closed. In 1891 both BGS and CGS were among the founding members of the Schools' Association of Victoria (SAV) which organised football and cricket matches amongst local private boys' schools. HC played its first inter school cricket match against CGS in 1892, with CGS being the victor. CGS has worn its traditional blue and white colours since the first years of its founding. Increasingly the use of urban paddocks as playing fields was frowned upon and from 1891 onwards all football matches had to be played on grounds recognised by the Victorian Football Association, (VFA). CGS was adjudged the Champion Football Team of the SAV competition in 1899.

systematic reading of the Word of God among the boys. He claimed that it was the only safe foundation for education.[34] Barnett was one of the founders and first secretary of the Church Missionary Association (now Society) in Victoria and like his predecessor, Rev. J.H. Davies, he also offered himself for overseas missionary service, eventually becoming the Anglican Archdeacon of Hong Kong. During this time, he founded St Stephen's College (for boys), Trinity College Canton and took a leading part in establishing Hong Kong University.[35]

Endnotes

1 Webber. H. *Years May Pass On. Caulfield Grammar School 1881–1981.* Wilke and Company Clayton. 1981. p10

2 Durie M. Caulfield Grammar School Founder's Day Address. 2008.

3 Durie M. ibid.

4 Davies. J.H. Personal Diary. Caulfield Grammar School Archives

5 *Argus.* Saturday 23 April 1881. p11

6 CGS Prospectus 1883. File 8 – Item 0039: WM Buntine Acquisition.

7 Webber H. ibid. p20

8 Wilkinson I. *The Fields at Play – 115 years of Sport at Caulfield Grammar School 1881–1996.* Playright Publishing. Sydney. p17

9 Wilkinson I. ibid. p17

10 Wilkinson I. ibid. p65

11 *Australasian.* 16 December 1882. p14

12 Stockings C.A.J. *The Torch and the Sword.* The History of the Army Cadet Movement in Australia. UNSW Press. Sydney, 2007. p7

13 *Victorian Government Gazette,* March 6, 1885, p710

14 *Argus.* 18 March 1884. p7

15 Stockings C. ibid. p33

16 Penrose. H. *Outside the Square. 125 Years of Caulfield Grammar School.* MUP. 2006. p7

17 Penrose. H. ibid. p7

18 Wilkinson. I. ibid. p28

19 *Australasian* No.1160. Vol. XLIV, Saturday 23 June 1888. p1374

20 *Australasian* No.1210. Vol. XLVI. Saturday 8 June 1889. p1191

21 *Australasian* No.1225. Vol. XLVII. Saturday 21 June 1889. p595

22 Durie M. ibid.

23 Webber H. ibid. p36

24 *The Cricket* October 1888 Vol.1 – No 2. Page 9

25 Webber. H. ibid. p44

26 Webber. H. ibid. p44

27 Webber. H. ibid. p44

28 Wilkinson I. ibid. p31

29 Ross P.M. Letter to D.J. Moran. 13 December 1992 (Held in the CGS Archives)

30 Adams T.L. 'Reminiscences.' *Caulfield Grammarian magazine.* June 1912. p96

31 Du Ve F.I. 'Reminiscences.' *Caulfield Grammarian magazine.* June 1913. p141

32 *Argus.* Greenwood. R. 1950.

33 Webber H. ibid. p40

34 Webber H. ibid. p42

35 *The Church of England Messenger.* December 26, 1930, p. 621 (Diocese of Melbourne, pub)

CHAPTER 2

CAULFIELD GRAMMAR
SCHOOL CONTINUES –
PREPARING SOLDIERS

As he departed for Hong Kong, Barnett undertook the sale of CGS to a man who for the next thirty-eight years would consolidate its educational and social position in Melbourne; one who would take the key foundation principles of Davies and Barnett as epitomized in the practices of sport, spiritual matters, academics and military training and ensure that they and their allied philosophies

WALTER BUNTINE

Walter Murray Buntine was the owner/ principal of CGS from 1896–1931. Previously a Resident Master in the CGS Boarding House under headmaster E J Barnett, the young Buntine had established Hawksburn Grammar school in 1893. In 1896 he amalgamated the two schools and immediately increased enrolments. He guided CGS through its early days in the Schools Association of Victoria (SAV) and saw its sporting prowess grow each year. Under his headship CGS teams won one swimming championship, 8 athletics championships, 7 tennis and cricket premierships and 20 football premierships: the latter including a run of 18 consecutive years. Buntine engineered the move from 'Halstead' to new and larger grounds and purpose-built facilities in Glen Eira Road, the current site. Buntine managed the

support for the effort and subsequent outpouring of grief at the involvement of many Grammarians in the First World War. He was actively involved in society away from CGS and was a lay reader of the Melbourne Anglican Diocese from 1920–49 and was the first President of the Church Missionary Society. He helped to found Ridley College (an Anglican theological college attached to the University of Melbourne) in 1909 and served on its Council from 1909 until 1952. Buntine also served on the Council of that university from 1933–37 and the Council of Public Education from 1935–38. CGS was formed into a public company in 1931 headed by a School Council. During his time as Headmaster, Buntine had seen CGS grow from a school of 120 to over 500 pupils.

became the bedrock of CGS into the future. That man was Walter Murray Buntine. Barnett wrote to the CGS parents at the time of the sale:

> I have no doubt that under the strict but kindly discipline of the new Headmaster, the School will continue to flourish even more than it has done in the past, and it will be a satisfaction to parents to know that the foundation stone of the School – a working spirit of Christianity based upon the daily lesson from the Word of God – will be left untouched.[1]

Walter Murray Buntine was born in 1866 near Rosedale in Victoria, grew up in Gippsland and became a boarder at Scotch College for the final two years of his education, before completing a Bachelor of Arts at the University of Melbourne. He became the founding principal and owner of the inner suburban Hawksburn Grammar School in 1893. However, by 1896 he had merged this enterprise with Caulfield Grammar School, when he took over the school's ownership. Buntine saw great advantages in merging Hawksburn Grammar School, with that of Caulfield Grammar School. The premises and grounds were more spacious and attractive than those at Hawksburn, and in addition Caulfield had premises specifically designed for a school with a spacious dwelling house set in five acres of land with another ten acres under its control. A school of twice the size could be conducted with a higher rate of profit on the Caulfield site rather than at landlocked Hawksburn.[2]

Buntine, the owner/principal of CGS from 1896–1932 had, as a schoolboy at Scotch College in 1884–85, come under the influence of one of these so-called 'Arnoldian missionaries',[3] the Scot Dr Alexander Morrison who was the Headmaster of Scotch College from 1857–1903. In coming to Australia, he was seen as one of the 'new breed of [educational] missionary come to influence the colonies.'[4] His influence can be borne out in the words of Sherington who stated:

> Eventually, in Australia as in Britain, athleticism, militarism, and imperialism became enmeshed. Playing fields prepared boys for the battlefields of empire.[5]

The manner in which Buntine would operate and organise CGS was detailed in part by Horace Webber, the author of the school's centenary history and a former staff member, who had worked for Buntine in the 1920s. Webber wrote that, after Buntine had experienced, as a matriculation pupil in 1884 and 1885, the Headmastership of Dr Alexander Morrison of Scotch College:

> He, like many boys who sat under Dr Morrison, had profound respect, not to say veneration, for the ways and person of this patriarch among schoolmasters

in Victoria. To Walter Buntine, it must have been axiomatic that the way Scotch College was run was the proper way to run a school.[6]

It was seen that the middle class in Australia accepted sport because of the values that it taught the young and reinforced for spectators of all ages; values such as loyalty, determination, unselfishness, and team spirit. Sport was a preparation, a training ground for something higher; he who succeeded in sport equipped himself to lead in business, politics, or the professions. Little wonder that sport attained such an exalted place in the private schools.[7]

Walter Murray Buntine was described as both a gentle man and a gentleman, of great integrity, with an honest face and someone who had faith in himself and in CGS. He had the walk of a (British) Guardsman and the complexion of an Englishman; he was an ardent Christian, an Empire man, and a true-blue conservative. His pupils were taught respect for the British Empire, the King and the King's representatives in Australia. He was always known as 'The Chief.'[8]

CGS horse drawn buggy in (circa) 1897. ('*Years May Pass On*'. P57)

In 1896, when Buntine took control of Caulfield Grammar School and merged it with his Hawksburn Grammar School, the total enrolments had reached one hundred and twelve; the first time in the school's history it had 'passed the century.' The original Hawksburn rolls have been lost, so it is unclear which boys had transferred to Caulfield, but among the new boys were three 'Boer War soldiers in waiting'.

Ormonde Winstanley Birch (1896–1898) was born on 14 May 1879. He was aged sixteen when he was enrolled at CGS on 14 July 1896 as student 580 and was living at 68 Sutherland Road Armadale.[9] He had previously attended Bendigo High School. His parents were John Edwards Octavius Birch and Clara Mary Cregoe who had four sons in total and one daughter. Birch was a good all-rounder at cricket and was named as a prominent batsman scoring 14 runs in the 1897 winning match against East Malvern Grammar School.[10] The following year his bowling prowess was noted in the winning match against Malvern Grammar School when he took figures of four wickets for fifteen runs.[11] The 1899 Speech Night program, when reporting on the Annual School Athletics Sports, stated that in the Old Boys' Cup in the 1-mile bicycle race, '*O Birch gained second place.*'

Hussey Burgh George Macartney (1886–1891) was born on 10 February 1875 and was the son of Emily and the Rev. Hussey Burgh Macartney who was

MALVERN GRAMMAR SCHOOL 1890–1961

Mr Charles McLean had founded Malvern Grammar School (MGS) in a house in Waverley Road, Malvern in 1890. With expanded growth the school grew to eighty students and moved to permanent premises in Kerferd Street Malvern in 1896. CGS played cricket matches against MGS in the years 1896–1898. John Davies, built the grand house Valentine's Mansions in Glen Iris in the 1890's. It had 40 rooms, a ballroom, a grand staircase and balconies, but Davies lost the property in the 1890's financial crash. In 1923 it was purchased by a group of former MGS students and the school took up residence there in 1924. Mr A.J. Marsden served as Headmaster from 1924–1955 and in 1929, MGS joined the Associated Grammar Schools Victoria (AGSV) sports competitions as a full member. The following year

of 1930, MGS shared the cricket premiership with CGS, but it would be ten years before MGS won its first cricket premiership outright, achieved off the last ball of its two-day match against Carey Baptist Grammar School in 1940. MGS won its first and only football premiership in 1944 when it shared the trophy with Ivanhoe Grammar School. In 1936 MGS published its first school magazine. The word 'Memorial' was added to its name after World War II in honour of those former students who had died during all conflict. Falling enrolments and a landlocked school site, meant MGS had little prospect of thriving or expanding and in 1960 CGS and MGS agreed to merge schools which meant MGS withdrew from the AGSV. From then onwards, Malvern boys completed their final school years at CGS.

the Dean of St Mary's Anglican Church Caulfield (now Oaktree Anglican) for thirty years from 1868–1898. Emily, a widow, had been previously married to Robert Kermode whose youngest son, Arthur Cotton Kermode (HBG's older stepbrother) was a CGS foundation student (no. 14 on the roll). Arthur attended Caulfield from 1881–1885 and started at the school on 31 April 1881, just six days after it opened.

Rev. Macartney was an instrumental figure in supporting and influencing Joseph Henry Davies, the founder of CGS to establish the school and later to embark upon his work on the missionary fields. Sarah Davies was Joseph's sister, and she had been sponsored by Rev. Macartney to serve as a missionary in India. Subsequently, Joseph would also serve at this mission before returning to Melbourne in 1878 and to establish CGS in 1881. Rev. Macartney remained a staunch supporter of CGS and presided over its early days at various public functions and Speech Nights. Macartney also gave Davies the responsibility of conducting the first services at the newly established mission church of St Clement's in Elsternwick. When the time came for Davies to sell CGS and move to the mission fields, Macartney helped to engineer the sale to Rev. E. J. Barnett as he had presided over the latter's ordination as a Deacon. Hussey Burgh George [Grammarian] was also the grandson of Hussey Burgh Macartney, the first Dean of Melbourne's St Paul's Cathedral.

George began his life at CGS in 1886 and given his families' church background, it was no surprise that with a High Average mark he won the Bible prize in Second Form, a prize he was to win again in 1888 and 1891. He appears to have been a good academic student and during his school years won prizes in Latin, Languages, Grammar, Arithmetic and Arts. In 1890 and a member of Form V, he demonstrated his sporting prowess when he won the Headmaster's Medal for Gymnastics. His work as a member of the newly founded school cadet unit was recognised in 1889, when he tied for first place with his classmate William Sargood for the Prize for Military Drill. Of note is that Sargood's father, Sir William Sargood, was at the time the Victorian Minister for Defence and had been instrumental in introducing the Cadet scheme to Victorian schools. Something can be told of George's character, even with the passage of time, as in 1891 his final year at CGS, he tied with a classmate to share *Mrs Were's Prize for Unselfish and Courteous Behaviour.*

Ronald Valentine Swanwater McPherson (1896–1898?) was born at St Arnaud on 10 August 1880, the son of William and Alice McPherson, who at the time of Ronald's enrolment at CGS in 1896 were living at 'Noorongong',

Mathoura Road, Toorak. He was noted as a very capable batsman and figured prominently in various cricket matches against local schools and in successive weeks in 1897, he was named among those who were successful 'with the willow.' In the match against Cumloden he top-scored with 19 runs,[12] the following week against Brighton Grammar School he made 67 runs before retiring voluntarily,[13] and with Ormonde Birch was equal top scorer with 14 runs in the winning team against East Malvern Grammar School.[14]

CGS sport historian Dr Ian Wilkinson surmised that in this year, CGS had not entered a team for the SAV premiership, but instead was allowing its first eleven cricket team to play 'friendly/social' matches against some second elevens from schools that had entered teams for the premiership such as Cumloden and Brighton.

Two years later the last 'Boer War' enlistee enrolled at CGS in the person of **George Frederick Roberts** (1898) who was the son of J H Roberts of Stevenson Street, Murchison, Victoria and was aged sixteen when he enrolled at CGS on 18 April 1897. He eventually entered the school as a boarder on 21 February 1898.

Moving ahead into the twentieth century, in 1961 Caulfield Grammar School had amalgamated with the Malvern Memorial Grammar School. This institution had been founded as Malvern Grammar School (MGS) in 1890 by Mr Charles McLean in a house in Waverley Road, East Malvern. The following year, with an expanded enrolment of fifty boys, he transferred to the School House of St John's Church, Malvern. Five years later in 1896, enrolments had grown to eighty, which resulted into a move to a purpose-built school premises at 16–18 Kerferd Street Malvern. Unfortunately, the original MGS school rolls cannot be found and hence being precise about the names of those former student that served in the Boer War can only be confirmed through other sources. Only four former MGS students who served in the Boer War have been identified, but it is possible that others remain yet to be found. Sadly, no details at all can be found of the 'Boer War' boys' schooldays, except for one newspaper report about a 'cricket ball throw.' Generalisations about the ethos of MGS and its general approach to education can be made in the context of the times and general approach of all educators of the late nineteenth century in Australia. MGS and CGS would have been similar in curriculum structure, academic thrust and the upholding of sporting pursuits as important to a boy's all-round education.

CGS 1899 Football Champions. (CGS Archives)

In conclusion, what factors had existed at Caulfield Grammar School to help to mould the minds and actions of the twenty-three young Caulfield Grammarians, and four Malvern Grammarians who would soon enlist for active service in the Boer War?

In Australia and especially Victoria in the later nineteenth century, most schools were privately run, as were Caulfield and Malvern Grammar Schools. As the historian Brown contends, '… the years preceding the turn of the century featured the consolidation of Australian versions of the British public school in Victoria and New South Wales. Central to the curricula of these schools was the powerful and addictive ideology of athleticism.'[15] Brown argued that '…the ideological foundations of athleticism had been advanced initially by a generation of British public-school masters who began to arrive in Australia during and after the 1860s. Their educational philosophy and practices were founded on the Arnoldian concept of a Christian education which was based on the development of character by means of 'playing the game' – literally and figuratively.[16]

Noted social historian Mangan saw a wider agenda in schools than merely just the attainment of commonly accepted educational goals to do with socialization, literacy and numeracy. He noted, 'In later Victorian and Edwardian Britain, to an extraordinary degree both sport and war were welded together into a fused expression of sublime middle-class heroic manhood with one as preparation for the other.'[17] Mangan contended that imperialism, militarism and athleticism in the last quarter of the nineteenth century became a revered secular trinity of the upper middle-class school. McIntosh agreed and wrote that, 'By the end of the century it was not the public-school system in general, but the playing fields that were associated with the imperial battlefields.'[18]

Other writers concurred and Meyer for instance stated that, 'the encouragement of competition was based on the moral grounds that games were a preparation for the battle of life and that they trained moral qualities, mainly respect for others, patient endurance, unflagging courage, self-reliance, self-control, vigour and decision of character.'[19] Some writers called this view of the prevailing religious ethos of the British Empire, 'Muscular Christianity.' This ethos was viewed in the following manner by one of its early proponents, the English writer Charles Kingsley, who saw through the medium of sport the potential for spiritual, moral and physical development.

> In the playing fields boys acquire virtues which no books can give them; not merely daring and endurance, but better still, temper, self-restraint, fairness, honour, unenvious approbation of another's success and all that 'give and take' of life which stand a man in good stead when he goes forth into the world and without which, indeed, his success is always maimed and partial.[20]

Muscular Christianity's call and chance to 'vindicate obligations 'and to carry out gallant service to the Empire at the outbreak of the Boer War in 1899 was built upon many long standing social and educational practices in Australia. Perhaps the best of these examples was in the spread of the Church of England, the adoption of sports such as cricket and rugby football, the copying of military uniforms, customs and practices in the army and the reproducing in the 'colonies' of the English Public School and its associated culture and trappings. The ethos of these schools was based on the English public-school tradition with its amalgam of learning, sport, military service in cadet units and loyalty to Queen, Country and Empire. In the context of Britain and implicitly the British Empire, Mangan in *Manufactured Masculinity* suggested that '... imperial masculinity was being methodically 'manufactured' by means of a cultural 'conveyor-belt' set up eventually throughout the empire with varying degrees of efficiency and with variable responses. Central to this was the Empire's influential public-school system, with its emphasis on muscular Christianity.'[21]

DEFENDING VICTORIA IN THE 19TH CENTURY

Between 1810 and 1870 a total of 24 British Army infantry regiments served in Australia, accompanied by units of the Royal Engineers and the Royal Artillery. Prior to this, volunteer units had been raised in each of the colonies since early times. Most of these were in Victoria, which at that time was the largest and most economically prosperous state, because of the gold rushes. When the last of the British troops were withdrawn in 1870, small government forces of infantry and artillery were raised in NSW and Victoria in the first steps towards creating permanent forces. In 1863 the *Volunteer Act* was passed in Victoria to raise a voluntary force of various arms including artillery and infantry. All mounted troops in the state became part of the Prince of Wales's Light Horse and in addition there was one company of engineers and seven of artillery. In 1870 the Victorian Permanent Artillery Corps of about 300 men was established to man the former British coastal fortifications, mainly at Queenscliff, Point Nepean and Swan Island. In 1884 a professional and paid force was established in Victoria, which incorporated the permanent Naval Forces, the Victorian Artillery and created the Victorian Militia Force. Two semi-professional units were established, the Victorian Mounted Rifles (VMR) in 1885, and the Victorian Rangers in 1888, mainly in country areas with large enlistments of local rifle club members. These clubs promoted the martial skills of marksmanship and rifle handling and often formed the basis of local militia units. On 31 December 1900, just before Federation, the strength of the Victorian colonial forces was shown to be 301 officers and 6034 troops.

The public-school boy of England [and Australia] had been trained to do his duty by the Empire and, because of the passion for games and athletic excellence which swept these institutions, it encouraged boys in particular to further develop those qualities needed to rule the state and Empire.[22] In time of war it then became an easy transition from loyalty to one's House at school to the same devotion directed towards one's regiment in the greater game of war. Boys were taught that success in war depended upon patriotism and military spirit and that preparation for war would strengthen 'manly virtue' and 'patriotic ardour.'[23] In the Empire's view, war was not an evil thing, but was the greatest of all games, a sport at which the 'home team,' the British Empire, invariably excelled.[24]

19th century sport in Melbourne

'The roots of Melbourne's distinctive sporting culture date from the earliest days of European settlement when the Melbourne Cricket Club, founded in 1838, acted as a focal point for a range of sporting activities, notably cricket and athletics. Melbourne also had an established racecourse with Flemington opening in 1840. The discovery of gold in 1851 had created a melting pot of immigrants. Various forms of folk football were played until 1858, when colonial-born Tom Wills, recently returned from Rugby School in England, provided the inspiration for a code that was initially known as Melbourne, then Victorian and finally Australian Rules Football, with the Melbourne Football Club (1858) and Geelong Football Club (1859) laying claim to being the oldest football clubs in the world. The first codification of rules took place in 1866, and the Victorian Football Association was formed in 1877, with some of the wealthier clubs seceding to form the Victorian Football League in 1896. The cult of athleticism that had its genesis in English public schools in the late 19th century had a pronounced effect on Melbourne schools. The rapid development of Australian Rules football, rowing, cricket and athletics, owed much not only to the efforts of enthusiastic schoolboys, but also to their various headmasters. Fervent about the character-building nature of sporting pursuits, the older private schools also promoted competitions such as the Head of the River rowing and organisations such as military cadet corps.'

At countless prize-givings, schoolboys were told that their schools were providing just that training in character and body which made future officers and leaders of men. Behind the creation of the officer, rifle and cadet corps was the new conception of the social citizen and the nation-in-arms.[25]

Indeed, Henry Newbolt's poem *Vitae Lampada,* written in 1892 and set in the grounds of his old school, Clifton College, was one of the best-known examples of combining sporting metaphors with an underlying 'call to arms.' With its catchcry *"Play up! play up! and play the game,"* it was one of many such vehicles used in the education of the young members of the Empire to teach them about the merits of a life of selfless commitment to duty. In Victoria, influential headmasters such as LA Adamson of Wesley College emphasized the importance of excellence in sport for his students and extolled the effects on character of training and preparation for military service.[26] He claimed that 'the British love of games had proved a magnificent asset to the Empire producing unselfish, devoted leaders, able to endure hardship and discomfort. Schoolboys might well continue with their games, even in times of war, because sport equipped them to take their places at the front if this was necessary.'[27]

Joseph Henry Davies, Ernest Judd Barnett and Walter Murray Buntine, the three headmasters of Caulfield Grammar School for its first twenty years of existence, fervently believed that their school should be a thoroughly Christian one. Apart from the paramountcy of academia, they all highly valued the importance of sport and organised games and made provision for them with suitable playing fields and admission to properly organised competition with like-minded schools. While the term 'Muscular Christianity' does not appear in any written form associated with CGS, its practices can be seen in the way the school's broader curriculum was conducted. As a first adopter of the school cadet movement in Victoria, CGS had bought into the movement that their near neighbour, parent of CGS students and cadet founder Lt. Col Frederick Sargood had sought, 'To bind together in one patriotic brotherhood the youth of this country so that, should occasion arise, they may be able in after years to defend their country with the most telling effect.'

Thomas Arnold of Rugby School fame would have seen in the students emerging from Australian schools, such as Caulfield and Malvern Grammar School, at the end of the 19th century, just those scholarly, Christian gentlemen he so desired to produce.

With the Second Anglo-Boer War about to break out in October 1899, all these underlying elements were about to combine to ensure that twenty-three Caulfield Grammarians and four Malvern Grammarians answered the call of 'Queen, Country and Empire.' For seven of them, to answer this call was to cost them their lives.

Endnotes

1   The Walter Murray Buntine Acquisition. FILE 2 – Item 017: Letter from Rev E. J. Barnett to CGS Parents; March 23rd, 1896 (CGS Archives)

2   Webber H. ibid. p58

3   In 1828 Thomas Arnold became the Headmaster of Rugby school in England. He propounded the view that the key role of a school should be the formation of character and that the boys should be trained not just to be scholars, but to be Christian gentlemen.

4   Brown D. ibid. p7 cited in Mangan. ibid. p393.

5   Sherington G. 'Athleticism in the Antipodes: The AAGPS of New South Wales,' *History of Education Review*, 12(2) (1983). p7

6   Webber H. ibid. p60

7   McKernan M. ibid. p95

8   Penrose H. ibid. p14

9   Gibson, J. Letter in reply to Robert Meyer. 8th June 2012.

10 *Australasian*. No. 1616. Vol. LXII. Saturday 20 March 1897. p580

11 *Australasian*. No.1670. Vol. LXIV. Saturday 3 April 1898. p747

12 *Australasian*. No.1614. Vol. LXII. Saturday 6 March 1897. p465

13 *Australasian*. No.1615. Vol. LXII. Saturday 13 March 1897. p516

14 *Australasian*. No.1670. Vol. LXIV. Saturday 3 April 1898. p747

15 Brown D. 'Athleticism in Australia: St Peter's College, Adelaide: A Case study in the diffusion of a Victorian Educational Ideology'. unpublished paper, University of Queensland, Dec. 1986, p7. cited in Mangan. ibid. p392

16 Brown D. ibid. p7. cited in Mangan. ibid. p393.

17 Mangan J.A. ibid. p, 15.

18 McIntosh P. *Physical Education in England since 1800* (London, 1968), p70. in Mangan ibid. p67.

19 Meyer. AR. CSKLS Annual General Meeting. Baylor University. 2010. *Journal of Sport and Social Issues.*

20 Haley. B. *The Healthy Body and Victorian Culture*. 1978. Cambridge: Harvard University Press. p119.

21 Mangan J.A. *Manufactured Masculinity: Making Imperial Manliness, Morality and Militarism.* (Sport in the Global Society – Historical Perspectives) Routledge. Abingdon. 2012. p9

22 James L. *The Rise and Fall of the British Empire*. London. 1982. p329

23 Steiner Z and Neilson K. *Britain and the Origins of the First World War* (2nd Ed) Palgrave MacMillan. Hampshire. UK. p168

24 McKernan M. *The Australian People and the Great War*. Nelson Australia. 1980. Melbourne. p1–2

25 Steiner Z. and Neilson K. ibid. pp167–8

26 Clements M.A. *Adamson, Lawrence Arthur (1860–1932)* Australian Dictionary of Biography, volume 7, (MUP), 1979

27 McKernan M. ibid. p101

PART TWO
THE BOER WAR
CONVENTIONAL
WAR PHASE

CHAPTER 3

Phase 1 – The Boers'
Early Success
October 1899 to December 1899

When Great Britain declared war on the Boer Republics on 5 October 1899 all the Australian colonies offered to send troops. In the parliaments of all the colonies speaker after speaker testified to their loyalty to empire and the crown. The Australian colonies were about to show the world their solidarity with the British Empire.[1]

By the time the New South Wales and Victorian units were ready to depart for the Transvaal enthusiasm ran high in Australia. On 3 October 1899 the [British] Colonial Office accepted the offer of colonial troops. It requested two units each from New South Wales and Victoria and one from each of the smaller colonies.[2]

Newspaper reports in Melbourne confirmed that, 'it can be said that Australia will be represented in the war in South Africa. The cable despatch from the Imperial authorities says that our men will be warmly welcomed and consequently we are put on our honour to despatch them.'[3] The same report wanted to pre-empt the conditions under which the colonial troops would serve in South Africa, as it pointed out that the British intended to draft the Australians to serve against the Boers by intermingling them with various British regiments. While some umbrage was taken at this potential development, the reality and value of Australian troops, yet to be tested on the battlefield, fighting alongside seasoned British soldiers was acknowledged:

> The Commander-in Chief may think that our amateur troops will have the better chance of distinguishing themselves, and that there will be less risk for himself also if they are made component parts of regiments that have had professional training and many of which have seen service in the field. The Australian born [people] in the colony are proud that they are of the British race, but there is a feeling that to establish our claim [as good fighting soldiers] our men must stand shoulder to shoulder with the British forces under the flag on the nations' battlefields, where manhood is tested, and empire is won or lost.[4]

Notwithstanding this potential disposition of colonial forces, young men all over

Australia rushed to join the various army regiments and contingents that were being raised by the Australian colonies for the South African conflict. *The Caulfield Grammar School Jubilee Book*, published in 1931, recalled the circumstances of the South African War [the Second Anglo-Boer War] only thirty years earlier. It explained that when it was announced that the Boer (farmers) of the Transvaal were preparing to invade the neighbouring state of Natal, an offer was made from Australia to send mounted infantry to assist the British forces in resisting the invasion. The *Jubilee Book* entry noted that:

> Not a few past students at the leading schools of Victoria readily joined up and proceeded to camp for training and in due course for active service in South Africa.[5]

The Australian contribution came forward in great waves of recruitment. The first contingents were raised by the colonies, which often drew heavily on men already serving in the militia of the various colonial forces. Two Victorian units, the Victorian Mounted Rifles (VMR) and the Victorian Mounted Infantry (VMI) who were called upon, were equipped, hastily received additional training and were ready for embarkation, just three weeks after hostilities in South Africa had begun. They became known as the 1st Victorian Contingent.

The Victorian (Volunteer) Mounted Rifles (VMR) had been gazetted as a Volunteer formation by General Order 559 of 2 December 1885, with the express aim of consolidating all the disparate cavalry units dotted around Victoria, into one co-ordinated and single administrative and military unit.[6] Headquartered in Melbourne, there were nine detachments spread around rural Victoria with H Detachment being quartered in Hamilton in the Western District. Other units were drawn from that town as well as nearby Branxholme, Casterton, Coleraine, Digby, Heywood and Merino. Almost immediately upon leaving CGS in 1896 and on his return home to Sandford, Tom Stock joined H Detachment of the VMR as Private no. 1421.

Tom and his older brother Duncan, who did not attend CGS, volunteered for the 1st Victorian Contingent and were accepted, Tom receiving the regimental number 89.

> The selection criteria were rigorous. Although there were far more names taken than were needed for the two companies (of the VMR), only the best men would go. Preferences would be given to those who were already serving with the militia.[7]

A tough medical examination also weeded out candidates with various ailments and poor teeth, dental care being scarce on the South African battlefields. Candidates also had to undergo a riding test and then undertake further training before embarkation. Generally, married men were not accepted, but some exceptions were made. Newspaper reports told of the ongoing training as the Victorian troops readied themselves for the fray.

> The [VMI] infantry settled down under canvas at Victoria Barracks on St Kilda Road and the [VMR] mounted men went into quarters at the Royal Show Grounds at Flemington. At either place has there been no time for idling. The mounted men had fresh horses to break in and to drill themselves into thorough fettle for what may be to them a tough campaign. They have had to train horses to stand the flash and noise of musketry and to take kindly to the halter. The infantry men have also had to put in some stiff drill, and to busy themselves with accoutrements, equipment, and other matters and both arms of the contingent have been doing some target practice.[8]

Tom Stock was joined in the contingent by Sergeant-Major Ernest Norman Coffey, a full-time soldier and the Drill Sergeant of the Caulfield Grammar School Cadet Unit.

Ernest Coffey had been born in New Zealand in 1862 to Johan and Mary Coffey and later migrated to Australia, where he married Louise Youlden in November 1885 and fathered four children between 1886 and 1898. He had served as a Sergeant with the 2nd Battalion Infantry Brigade in the Victorian Forces where this unit won the Brassey Medal for drill and rifle shooting.[9]

The Caulfield Grammar School Cadet Unit had been formed in 1885 and was only the fifth such unit in the state at that time. Coffey was a permanent member of the Staff of the Militia and was based at the Orderly Room in Richmond and also undertook the duties of the CGS Cadet Unit Drill Sergeant. As with

Company Sergeant-Major Ernest Norman Coffey a member of the 1st VMI and the Drill Instructor of the Caulfield Grammar School Cadet Unit. (*The Leader* supplement)[10]

many other soldiers who were members of militia or regular forces, Ernest Coffey enlisted as a Company Sergeant-Major (CSM) in the 1st Victorian Mounted Infantry with the Regimental Number 1. In June 1900 another permanent army member named Downey was seconded from the now Royal Australian Artillery (RAA) and temporarily appointed as a Permanent Staff of the Militia Infantry (Victorian Rangers):

> … to perform the duties of Company Sergeant-Major during the absence from the colony, on duty, of CSM E.N. Coffey.

Saturday 28 October 1899 saw Victoria's first contingent, the Victorian Mounted Rifles (VMR) and the Victorian Mounted Infantry (VMI), depart with great fanfare for South Africa amid the acclamation of the citizens of Melbourne. *The Argus* newspaper noted:

> The 1st Contingent could justifiably claim to be the elite of Victoria's militia. After all, hadn't they been selected as such?' It is estimated that some 250,000 well-wishers were on the streets when the 1st Contingent marched on its way through Melbourne to the Port Melbourne Pier.[11]

Melbourne's population in 1899 was about 450,000, so the turnout was the largest ever seen to date in the city's history. The *CGS Jubilee Book* noted that:

> It seemed that all Melbourne turned out to see the first detachment of these Victorian soldiers as they rode four abreast to the waiting steamer *Medic* which was to convey them to the scene of hostilities.[12]

A contemporary account tells of their departure for South Africa:

> *The Argus* newspaper estimated that some 250,000 well-wishers were on the streets of Melbourne when the 1st Contingent [including Tom Stock and Norman Coffey] marched on its way to the Port Melbourne pier. Caught up in the

Tom Stock (1893–96) was the first Caulfield Grammarian soldier to serve overseas and sadly, the first to lose his life in action.

excitement of the moment, the newspaper declared that people had come to see big fine strong fellows, the flower of the colony's youth and strength, the first offering, the very first the colony had ever offered to her motherland and the crowd cheered and shouted as they passed along. When the mother country fights … we her children are eager and anxious to stand at her side.[13]

The first Victorian Contingent march through Melbourne on Saturday 28 October 1899. (angloboerwar.com)

Such was the patriotic fervour aroused in Melbourne that, *The Leader* newspaper printed a special souvenir edition on Saturday 28 October 1899 marking, 'The Departure of the Victorian Contingent to the Transvaal'. The 16-page publication outlined such matters as the cause of the war, included biographies and photographs of the British Commanders, as well as photographs of the main Boer commanders. Also included were details of the parade to the ship and most surprisingly, the names and 'passport type' photographs of all the Victorian men sailing on the *Medic* for South Africa. Private Tom Stock was photo 2 in the newspaper and was shown amongst the Mounted Rifles Units, whilst Sergeant Major Ernest Norman Coffey, the CGS Cadet Unit Drill Instructor, was depicted amongst the (Victorian Rangers) Infantry Unit. Remarks about their departure were made at the CGS Speech Night in December 1899:

Tom Stock, who was a boarder in 1896, along with our Drill Sergeant, (Coffey) is representing the school at the war in South Africa.[14]

The four-week voyage of the *Medic* saw the troops stop at Albany in Western Australia, where they paraded through the streets and attended banquets hosted by the Mayor and Councillors, before reaching their destination of Cape Town in South Africa on 26 November 1899. Some of Tom's letters home to his sister were published in the *Casterton News* newspaper and he described some of his experiences which greatly informed the local readers about the War.

> *Medic* – November 21. 'We have had a glorious voyage up to date and it has been wonderfully calm. But there has been a great deal of sickness since we started, and I had the influenza myself. The day after we left Albany every soul on board was vaccinated and it played the dickens with some of us. Both Duncan and I have had very bad arms, mine twice the size it ought to be, and it is very sore. We have been having sport onboard today, but there was scarcely enough room, as the horses take up all the deck. The food generally is very poor.[15]
>
> *Medic* – November 25. 'We are in for heavy weather now and it has been very rough last night and this morning with seas breaking over the ship every five minutes. The horses do look very wretched this morning as some of them are standing in six inches of water and it has been so rough, we could not feed them properly. We gave them hay, but it soon got saturated with seawater and they would not look at it.[16]

Finally, on Monday 27th November, the Victorians docked at Cape Town and Tom hinted at seeing his first action against the enemy when he wrote:

We finally got ashore on Monday morning and proceeded directly to our camp, 4 miles out of town and there are thousands camped here. We are to start for the front tomorrow, and we are all equipped with the new Lee-Metford rifles, and I suppose by the time you get this, we will have smelt [gun] powder.[17]

The largest forces in South Africa equipped with many of these new rifles were the regular British Army and regiments and former Caulfield Grammarian Hussey Burgh George

Lt. H.B.G. Macartney (https://www.youtube.com/watch?v=tBX37tgXM08)

Macartney (1896–92) served in one such unit. Following six years education at CGS, George had accompanied his father to England in early 1893 and undertook further studies at St Leonard's-On-Sea at Sussex, before matriculating and being enrolled at Cambridge University's Trinity College in September 1893. He graduated with a B.A. in 1896 and was looking to enter holy orders but abandoned this intention and enlisted as a 2nd Lieutenant in the 2nd Battalion of the (London Regiment) in 1898. This line infantry 2nd Battalion had existed in various forms since 1685 and had served in the American War of Independence, the Napoleonic Wars and the Crimean War. In November 1899 soon after the

BOER COMMANDERS

Paul Kruger or 'Oom Pauk' (Uncle Paul) was elected as the third (and last) President of the independent Transvaal Republic in 1883. He had almost no formal education but was an astute and intelligent leader. When Britain issued an ultimatum to Kruger in October 1899, he responded by launching a pre-emptive strike against them. Eventually he went into exile in Europe and died in 1904. **Martinus Steyn** was the last President of the Orange Free State and when the tide of the conflict turned against the Boer, was instrumental in coordinating the guerilla war against the British. **General Koos De La Rey** was a South African military officer who served with the Boer forces as a General and had some success against the British in actions at Graspan, Modder River and Magersfontein. He eventually commanded the guerilla war in the Western Transvaal for the last two years of the war. **General Louis Botha**

was the man chiefly responsible for the Boer's stubborn and very effective guerilla campaign against the British. He scored victories against their forces at the Battles of Colenso and Spion Kop. It was only when the British doggedly pursued a scorched earth policy against the Boer farms that resistance was halted. **General Christiaan R De Wet** had fought bravely at the Battle of Majuba Hill and in the early days of the Boer War commanded many successful counter attacks against the British. Because of his daring raids and attacks he was the most feared, legendary and well-known Boer leader. **Jan Smuts** was a lawyer who practised in Pretoria, the capital of the South African Republic. He became Paul Kruger's 'eyes and ears' in that city and handled propaganda and communications for the Boer. From mid-1900 onwards he served with Koos de La Rey's guerilla forces in Western Transvaal.

outbreak of hostilities, the battalion had been ordered to South Africa. In very early 1900 the battalion was absorbed into a larger brigade and became part of the force attempting to break the Boer siege on the city of Ladysmith in Natal. Macartney was involved in the various actions involved in trying to lift the siege, including one which was to cost him dearly.

The period between the outbreak of hostilities in early October 1899 and December of that year was a time when the British armies, mainly infantry were defeated or besieged by highly mobile Boer mounted troops. In early October

BRITISH COMMANDERS

General Sir Redvers Buller VC was the British Army officer who was the Commander-in-Chief (C-in-C) of all British Forces in South Africa in the early months of the Boer War. He was a popular military leader, but suffered numerous setbacks and defeats and was replaced by Lord Roberts in January 1900. **Field Marshal** (later to be Lord) **Frederick 'Bobs' Roberts VC,** was one of the most successful British military commanders of his time. Upon becoming Commander-in-Chief, he vigorously regained the initiative for the British and by December 1900 had forced the Boers to either submit or return to their farms. He introduced the controversial practises of burning Boer farms and then housing the families, without their menfolk, in concentration camps. On 12 December 1900, Roberts handed command to **Field Marshal Herbert** (later Lord) **Kitchener,** who expanded the successful practices devised by Roberts, even though there were appalling death rates in the concentration camps. His conduct of the war was once described as a 'scorched earth policy' where Boer farms and houses were burned to the ground and wells were poisoned, all in the name of denying sustenance to the Boer guerillas. At war's end he returned to London as a conquering hero. **Major-General John French** commanded the 1st Cavalry Brigade during the Boer War and his troops were heavily involved in relieving the Siege of Kimberly. His cavalry forces charged Boer positions at Elandslaagte with deadly results and were key to the success of the action at Diamond Hill. French became C-in-C in June 1901. **Lt-General Paul Methuen** expelled the Boers from Belmont and Graspan before being slightly wounded at the Battle of the Modder River. He suffered a defeat at the Battle of Magersfontein and Tweebosch. He was later captured by the Boer in March 1902, but was released by Koos de la Rey.

1899, the Boers invaded the Cape Colony having declared war on the British. Aside from scattered war parties, their main efforts were centred around the western border town sieges of Mafeking and Kimberley. During this time, between 10 and 17 December 1899, (Black Week) the British army suffered three terrible defeats from the Boer irregulars at the Battles of Stormberg, Magersfontein and Colenso with a total of 2,776 British soldiers becoming casualties. Macartney's unit fought very bravely at the Battle of Colenso. Eventually in February 1900, Kimberley was relieved and then in May, the town of Mafeking. The time between 13 October 1899 and 28 February 1900 also saw the invasion of the British colony of Natal by troops from the South African Republic (Transvaal) and the Orange Free State. The Boers severed the train link to Durban the capital, defeated the first British effort to relieve Ladysmith at Colenso and besieged Ladysmith for 188 days.

Tom and Duncan Stock's journey in South Africa really began on 1st December, when the troops boarded the train at Cape Town headed north for the De Aar region, which was then seen as the seat of the war. Large crowds had gathered to wave them goodbye, and they travelled to Orange River station where they left the train and marched 20 miles or so to Belmont as described in one of Tom's letters.

> The Cape people gave us a great send off and presented us with papers, cigarettes and beer. One fellow gave me a bottle of Three Star brandy which I emptied into my water bottle. It is an awfully barren country that we have to pass through with not a blade of vegetation to be seen anywhere. Some of the places have not seen rain for three years and the country seems to be one mass of rocks and mountains.[18]

He was a keen observer of his surroundings and in a 2 December letter, took note of the people and circumstances he encountered.

> We are passing trains with prisoners and wounded on board all day. The carriages for the wounded are splendidly fitted up with a good bunk for each man who is sent back if they have even the slightest wound. The Boer prisoners are a hangdog looking lot. We are the last of the colonial troops going up to the front line and all our fellows seem very eager to get there.[19]

When he arrived at Orange River, he wrote that:

> It is one of the dustiest places that can be imagined, and it is a very big camp with between 5–6,000 troops here. We are within touch of the Boers who are only a

few miles away and we are between them and their main body, so we might have a scrape with them at any time.[20]

Unusual encounters also caught his attention:

The nights are very cold. There are plenty of ostriches about here and I must try and get some feathers to take back with me as one of the fellows was out today and picked a beauty.[21]

But military matters were his main concern, and he reported on some preparations.

I would like to be in the advance to Kimberley as the regiment that gets there first will get £5 each from Cecil Rhodes. We are evidently in for a bit of hot work tomorrow, as we all must carry 150 rounds of ammunition with us, and we also carry a day's worth of food for ourselves and the horses.[22]

WEAPONS, TACTICS AND UNIFORMS

At the beginning of the Boer War, the British Army still relied on the tactics of past wars and often used single shot firearms fired in controlled volleys and fighting in close formation. They soon adopted the new Lee-Metford .303 bolt action rifle which had a ten-shot magazine, while the Martini-Henry .45 single-shot rifle was renowned for its accuracy. The Boers were generally equipped with German Mauser bolt-action five shot magazine rifles which fired smokeless ammunition, ideal for guerilla warfare. In addition, both sides employed early versions of machine guns, with the British using .303 Maxim guns, while the Boer used a gas-operated Lewis gun. With strong field craft skills and high mobility, born from lifetimes as farmers on horseback, the Boer proved to be a formidable natural mounted infantry, adept at living off the land. In addition, the Boer wore their everyday civilian clothing without badges of rank or other embellishment. The continually accurate and deadly Boer marksmanship meant that British officers soon realised that shiny badges of rank and the carrying of swords on the battleground also made them an easy target. By the end of the war the uniform of choice was a slouch type hat, with a dull drab tunic and trousers. It was also the troops of the British Empire, especially those from Australia, who were not attached to past British military practices, that demonstrated that it was more sensible in South Africa to use their horses for rapid movement and to advance in loose formation, rather than a formal massed infantry advance in tight ranks.

As the VMR moved further into Boer territory their general work as soldiers changed, as they encountered a type of warfare they had never encountered before, in the form of guerrilla warfare. The VMR and their associated mounted troops were used for patrolling the surrounding countryside, and they came to rely on the big guns of the British artillery batteries to dislodge the Boer from their entrenched positions. In time, the VMR slept all day and then undertook patrols at night to seek out the elusive Boer. But Tom wrote home about the reality of what he had encountered.

> Thank you for the copies of the *Casterton News*, which we were very pleased to receive. People made a great mistake when they thought that the war would be over in a short time, as it will last a long while yet as every Dutchmen in South Africa is fighting for the Transvaal.[23]

But at the same time Tom had his opinions about his Boer foe and expressed them in his letters home.

> The Boers have not touched the [railway] line since we have been here and in fact, we have not seen a Boer hardly lately and the few we have seen have run away as if the Devil was after them. They say they are very frightened of mounted men, and they believe that we are crack shots and horsemen and would have a rough time if they ran into us. The Boers are terribly afraid of steel [bayonets] and fall on their knees and call for mercy.[24]

In mid-December some of the VMR had moved north to the rail siding of Enslin and Tom noted of the location and his conditions.

> The weather is not very good. It is terribly hot here in the daytime, there is no shade of any sort to get under and the tents are like furnaces. There are an awful lot of deserted farms around here as the Boers cleared all the British farmers out and then when the English troops arrived, they cleared all the Dutch residents [Boers] out. The other day we were out and came to a deserted farmhouse and got ten geese, some onions and a cabbage. We cooked and stuffed two of the geese with onions and ate them and they were very good.[25]

On 27 December Tom wrote home about how he had spent Christmas, his food in general and his encounters with some British troops.

> It is very hard to find anything to write about in this dead and alive hole as it is the same thing day after day. We had a very nice Christmas here, although we were to get all sorts of nice things for Christmas dinner, we did not get anything special,

the goods not arriving in time. All we got was soup, beef and potatoes. No plum puddings as I believe they went astray somewhere down the line. They say we are to get them on New Year's Day. The Gordons [a British regiment] get jams, milk, fish and other things sent up by train and sell what they want to us; but we must pay well for it. All of us H Company fellows are together in the same tent, and we now have a box full of milk, jam and fish in our tent now, so we are not so badly off for now. The Gordons are a fine lot of fellows and very pleasant to speak to and were in India for 18 years before they came here, so they have something to talk about. They are very fond of their beer, but I think it is very bad and almost as thick as gruel. We are allowed a pint of rum to each tent once a week and as it is a long way over-proof, it is almost as good as two pints. It doesn't take too much to put a man on his ear. They can keep the rum for my part as I can't bear the smell of the stuff.[26]

A few days after Christmas, saw Tom out on patrol looking for the Boer and ensuring that communications and supply lines were safe.

I was out patrolling the railway line last night and I was nearly frozen as the nights are bitterly cold, whilst during the day the sun shines like a furnace. It is not very pleasant work out at night when there is no moon. There are so many stones and holes in the ground made by moles and other animals that one must be very careful where he is going or else his horse will come over. My horse is very frightened at night, and he snorts and jumps away from everything he sees. Both Duncan and I are keeping in splendid health now and I have never felt better in my life and am getting quite used to it and rather like it.[27]

A few days later, on New Year's Eve Tom wrote home to his sister about a positive change in conditions.

We have just received the mail, and I got your letter. It was awfully good to hear from home as it seemed an age since I had heard from you before. I also received the *Casterton News*, which was very acceptable as we get very little to read. Dossie Bolton also sent me the *Australasian*, the *Weekly Times*, and a very pretty Christmas card and I was the only one in the whole division who received a card. We are doing very well in this camp now and can buy almost anything we want, as rail trucks of provisions come up every day and they are not allowed to overcharge like they did at first. We buy Quaker oats in 21lb packets and with the boiler in our tent make porridge every morning.[28]

Aside from these conditions for the Australians, the first few months of the

BLACK WEEK

During the week of Sunday 10 December to Sunday 17 December 1899, the British Army suffered three disastrous defeats at the hands of the Boer Republics. In total **2,776** British soldiers were killed, wounded or captured during this week. The British had assumed they would easily defeat the irregular Boer troops, but had to drastically re-think their strategies after this terrible week. **10 December – Stormberg**. The Boers had seized the important railway junction at Stormberg in the Cape Colony and so the British sent troops to attempt to recapture it. With no reconnaissance and little rest, the tired British troops stormed Boer positions on a hill but were mistakenly shelled by their own side. Forced to retreat, the fleeing British troops were attacked by mounted Boer horseman, with nearly 700 soldiers captured. **11 December –**

Magersfontein. At this important rail crossing, Lord Methuen had failed to order enough reconnaissance and was not aware of superior Boer forces in the area. Most Boer survived the initial artillery barrage, and the advancing British troops suffered heavy casualties from accurate Boer rifle fire. Almost 1000 British soldiers were killed or wounded. **15 December – Colenso.** General Buller VC, wanted to relieve the city of Ladysmith as soon as he could and mounted a frontal assault on the key river crossing at the village of Colenso in Natal. Poor early reconnaissance and planning, led to British soldiers being subjected to deadly Boer artillery and rifle fire and they lost 143 killed and 756 wounded. Among this last group was Caulfield Grammarian **George Macartney**, a lieutenant with the 2nd Royal Fusiliers.

South African conflict had been very costly for the British Army in men killed and material and territory conceded. The 'Black Week' of 10–17 December 1899 opened the eyes of the British, who had thought that the Boer irregular troops would be no match for their own professional army. Swiftly the British government undertook many necessary changes to some high-ranking military leadership, arranged for better mobilisation of troops across the empire and made quick progress to modernise the army's equipment and tactics. Many infantry units were better deployed or became mounted regiments equipped with rapid-firing rifles rather than swords and lances. Machine guns became an integral part of the fighting unit's arsenal. Perhaps the greatest change that was instituted was to realise that the Boer fought on horseback and that if the British were to match them, they must employ similar tactics. At this stage of the war, help was certainly on its way from Australia, especially in the form of mounted regiments with many of the riders experienced and brave horsemen. Tom Stock and Ernest Coffey were already in the field. Among the thousands of young Australians who arrived in South Africa in 1900 to provide the British with much needed support and assistance were fourteen Caulfield and Malvern Grammarians. Their units of origin were spread across four Australian colonies, and one even travelled independently to the conflict and enlisted in a South African regiment. The year of 1900, however, would bring not only a change of tactics and fortunes for the British forces, but also tragedy for three Caulfield Grammarian 'Soldiers of the Queen.'

Endnotes

1 Clark. M. *A History of Australia. The People Make Laws.* Volume V. MUP. Carlton. 1987. p169.

2 Clark. M. ibid. p170.

3 *Argus* Friday 6 October 1899. p4

4 *Argus* Friday 6 October 1899. p4

5 *Caulfield Grammar School Jubilee Book.* 1881–1931. p39

6 Victorian (Volunteer) Mounted Rifles http://alh-research.tripod.com/LightHorse/index.blog?topic_id=1113235

7 Droogleever. R. *Colonel Tom's Boys. Being the Regimental History of the 1st and 2nd Victorian Contingents in the Boer War.* 2013. PrintBooks, South Melbourne. p18

8 *Weekly Times.* Saturday 28 October 1899. p13

9 In 1896 the Governor of Victoria, Lord Thomas Brassey, an enthusiastic supporter of the Local Forces introduced a competition known as the Brassey Field Firing and Marching Competition. Coffey's Unit won the inaugural competition in 1896 and went on to win it another three times.

10 The Melbourne publication *The Leader* published a sixteen-page supplement on 28 October 1899, the day of the departure of the first Victorian Contingent to South Africa. It contained a background to the cause of the Boer War, short biographies and photos of the key British and Boer military personnel, details of the farewell parade and a map of the area around Ladysmith. The most remarkable features of the supplement are the 'passport-type' photos of every Victorian soldier in the contingent accompanied by a nominal roll with their rank, serial number and town of residence.

11 Droogleever. R. ibid. p22

12 *Caulfield Grammar School Jubilee Book.* ibid. p39

13 Droogleever. R. ibid. pp22–23

14 Caulfield Grammar School Speech Night program. 1899. p3

15 Wilmot. P. (Ed) Tom Stock Letters published in the *Casterton News.* Quoted from *The Second Harvest – The Writings of Tom and Duncan Stock, two Victorian brothers in the Boer War, 1899–1902.* 2000. p5

16 Wilmot. P. ibid. p5

17 Wilmot. P. ibid. p6

18 Wilmot. P. ibid. p6

19 Wilmot. P. ibid. pp6–7

20 Wilmot. P. ibid. p7

21 Wilmot. P. ibid. p7

22 Wilmot. P. ibid. p7

23 Wilmot. P. ibid. p10

24 Wilmot. P. ibid. pp9–10

25 Wilmot. P. ibid. pp9–10

26 Wilmot. P. ibid. pp10–11

27 Wilmot. P. ibid. pp11–12

28 Wilmot. P. ibid. p12

CHAPTER 4

PHASE 2 (PART 1) – THE BRITISH REGAIN THE ADVANTAGE
JANUARY TO SEPTEMBER 1900

The second phase of the Boer War from about December 1899 to September 1900, saw several major British counter-offensives, resulting in the capture of most of the major towns and cities of South Africa. The troops that arrived from Australia during this time can be divided into two major categories. The first group were the 'Bushmen' contingents which had been recruited from more diverse sources than the first former 'militia' based contingents. These units had often been paid for by public subscription or wealthy individuals and they generally arrived at the conflict between November 1899 and March 1900. The second group were the 'Imperial Bushmen Contingents' which in many ways were very much like the preceding contingents, but were being paid for by the British government. They generally arrived in South Africa between April and June 1900.

Between 3 January and 18 May 1900, the large British forces, now approaching 180,000 in number, retaliated against the Boers who at their peak numbered about 88,000 men. With an influx of other British army units and augmented by many colonial units, including those with Caulfield and Malvern Grammarian soldiers, the British chased the Boer forces out of the Cape Colony and Natal. In addition, the weight of British numbers enabled them to annex the Orange Free State in May and then move operations further north and capture the Transvaal in late October. The Boers were now faced with much superior British numbers and were forced to move to a strategy of guerilla warfare, which extended the conflict for another two years. For the rest of the war the practice for both sides were minimal set piece battles, replaced by skirmishes and raids of various sizes, as well as interminable patrolling by the British to harass and neutralise the Boer commandos.

By 10 January 1900, the 1st Victorian Contingent had been reinforced by troops from all the Australian colonies and the rest of the Victorians had also moved from Belmont to the rail siding at Enslin, about 18 miles south of the Modder River. There they continued their relentless patrolling in the hot and

BOER RIFLES AND AUSTRALIAN HORSES

The Boer War was noted for several innovations. The Boer used Mauser rifles accurate to 2000 metres with smokeless ammunition, to target their enemy at long range without the British seeing the origin of the shots. For the British side having the larger forces, the war transitioned from a series of set battles using infantry, to a mobile guerilla fight across a wide country. The combat became one of mounted mobility by a larger force (British) against a better armed and skilful enemy (Boer) who fought on their home ground. Horse mounted soldiers who used their horses for mobility but then fought on foot, became the most capable combat arm. British cavalry units still carried their shock weapons, but generally left their lances and swords with their horses and fought on foot with their rifles. Any infantry units, where possible, soon transitioned into mounted units, as with the VMI. Australian soldiers with their horsemanship, bushcraft and shooting skills honed in very similar 'rural' conditions to that of South Africa, were very effective and highly regarded soldiers. The emergence and importance of the military horse in this conflict was paramount for two main reasons. Firstly, the fighting took place on a battlefield ranging over thousands of kilometres, and secondly the nature of the 'guerilla warfare' used by the Boer, made it essential for both sides to be mobile for quick transport, as opposed to infantry fighting. Australian forces generally took their own 'Waler' horses to South Africa. With their origins reaching back to the First Fleet, these horses had cross-bred with other horses, resulting in a type of strong bush horse suited to the Australian conditions and climate. Because of their strength, they were used by early explorers, stockmen, farmers and especially the military.
(Source. bwm.org.au/course.php)

dusty conditions. A month later, on 10 February, the Victorians were relocated to two major camps, the first of which was Windmill. This was around three miles east of Colesberg, where the South Australians and the Victorian Mounted Rifles under McLeish were stationed. About a mile from Colesberg, was the second camp, Kloof Camp, where the Victorian Mounted Infantry and the 2nd Battalion of the Duke of Edinburgh's (Wiltshire) Regiment were garrisoned. On that day, two detachments of the VMR occupied outposts at Bastard's Nek and Hobkirk's Farm.[1]

Tom Stock wrote on 14 January, from the rail siding at Enslin camp, that he was still stuck there and by all accounts was likely to remain there for some time.

> We have done absolutely nothing yet but patrol the country around about. There
> are supposed to be about 2,000 Boers just 20 miles from here and we went with
> the 9th Lancers and two batteries of field artillery to shift them, but we could not
> see anyone. We were highly complimented by the officer in charge for the way we
> scouted the country for the rest of the troops. War is altogether a different thing to
> what I thought it to be, as we are all as contented as can be here with nothing to
> trouble one. You might as well send us the *Australasian* every week as not having
> anything to read is the worst part of the business.[2]

But despite Tom's 'contented' approach to warfare, its reality was not too far
away, as he wrote on 14 January 1900 of an experience, he had undergone the
day before.

> I had an experience yesterday that I won't forget in a hurry and that was burying
> a man who had been dead since the battle of Belmont [on 23 November 1899].
> He was lying away out on the veldt and had evidently been missed when they
> were collecting the dead. A patrol of our fellows found him the day before. He
> was a member of the York Light Infantry and had been shot through the head and
> stomach and was in a very decomposed state. It nearly made me sick to look at
> him. There was a half-written letter to his sister in his pocket and we buried him
> where he lay and stacked some stones over the top of him before the Gordon's
> chaplain came and read the burial service over him.[3]

Tom wrote again on 1 and 2 February in a letter in two parts which detailed his
soldiering activities in late January in pursuit of the Boers in country to the west
of Enslin.

> On January 22nd we camped at a farmhouse a few miles from Douglas and sent
> our picquets and patrols to watch the surrounding countryside. Towards sundown
> one of the picquets noticed two parties of Boers riding towards them. Our men
> immediately retired to gain a better position, and the Boers opened fire on them.
> Our men dismounted and returned the fire and the Boer withdrew. I was in a
> party that was going out to relieve the picquet, when we heard the shots, and we
> galloped towards them and saw them galloping towards us for all they were worth.
> For the first time, one of our men was wounded in the arm but managed to return
> to our lines. We thought we were going to have a lively time of it, but the Boers
> must have retired.[4]

Sergeant-Major Ernest Coffey, who had sailed with Tom Stock from Melbourne, was attached to the Victorian Mounted Infantry (VMI) and wrote home on 18 February of his time in South Africa in late January and early February.

> We left Enslin by road on 30th January for Belmont. Left again next day by rail for Naauwpoort and arrived there at 3.30 next morning. After passing a terribly cold night in open goods trucks, without any covering, we got fitted out with horses, and left for Rensburg by rail on 3rd February, arriving there at 10.30 same night, bivouacked with our horses till daylight, and then pitched our camp. Our horses were handed over to the New South Wales and West Australian companies to complete their number, as there were still 150 horses short in the regiment, and fresh ones were to come on for us.[5]

The British Army had rejected the use of a new type of rapid-fire field artillery due to its 'light firing capacity' compared to their current field artillery pieces. The Boer forces, however, realised that the rapid rate of fire and the ease of mobility of the new Maxim [pom-pom] guns would be of great advantage in the new type of mobile warfare being fought in South Africa. Coffey recounted his first encounter with the new enemy weapon.

> On the 5th inst. 100 of our company on foot left Rensburg for Kloof camp (to the north-west of Colesberg) and arrived there same night. The Victorians had to pass over a dangerous piece of ground on the route — ground that was covered by the Boers from Colesberg. Our men had 17 [pom-pom] shells put after them but were lucky enough to escape a hit. The men were well extended in open order, and the first few shells that came after them were badly placed, which caused our men to laugh and chaff over the inferior marksmanship, but each successive shot came closer, and the last one that was fired at them, as they were disappearing behind a kopje, burst 10 yards from them, after passing just over their heads. These shells were fired from a Maxim-Vickers automatic gun, which fires five or seven 1 lb. shells continuously; our men call them the pom-poms, or 10 a penny.[6]
>
> The Victorians stood their christening well, and there was no appearance of funk amongst them.[7]

In the meantime, the VMR, including Tom Stock, had left Rensburg (five miles from Kloof camp) the day before the VMI left. The column of some VMI and VMR now also comprised 100 of the New South Wales company mounted at Kloof, 100 of the West Australian company at Slingersfontein, 40 of the Tasmanian company at Jassfontein, and the remainder of the VMI regiment at Rensburg.

Tom Stock's photo as it appeared in *The Leader* newspaper supplement on the day of the embarkation of the troops from Melbourne in 1899. (*The Leader*)

In what turned out to be his last letter home, Tom reflected on his previous encounter with the Boer at Douglas as well as reflecting on a recent gift.

I suppose you have read all sorts of reports in the papers about us, but I am afraid none of them have any truth in them. The affair at Douglas was the most exciting and that was not much. I don't think the war will last much longer, as the Boers are beginning to fight starvation as well as us and that is the deadliest enemy to fight of all. Once they go down, it will be all of a heap. We received our chocolates that were presented to us by the Queen this morning. They are served up in tin boxes with the Queen's head on them and some of the men are selling the boxes for as much as £5, so they are a curiosity.[8]

From the town of Naauwpoort, on 2 February, he also reported on troop movements and his role in the next phase of VMR's involvement in the conflict.

We arrived here by train last night and this is rather a nice camp and a good-sized town which is the head depot for the region. We are to proceed north-east at 2.00pm this afternoon to join General French at Colesberg which is only 30 miles from here. He is supposed to have the Boers hemmed in and wants more cavalry to help him, so we will have something to do at last. I received the *Casterton News* this morning, which was very acceptable. I must conclude this letter as I must go and water my horse. Hoping you are all well as this leaves Duncan and me at present. Goodbye. Tom Stock.[9]

Sadly, this was to be Tom's last letter home and, as others pointed out later, it was the only letter he wrote home that he signed off with 'Goodbye' rather than his usual sign off.

The events that led to Tom Stock's death in battle are told in an account by Dr Chris Coulthard-Clark in his book, *The Encyclopedia of Australia's Battles.*

Pink Hill, a famous action during the Second South African War, fought between Australian (and British) troops and a Boer force on 12 February 1900, some thirteen kilometres north-west of Colesberg in central Cape Colony. The engagement occurred after the Boers, realising the weakness of the column of 6,600 men under Major General R.A.P. Clements which opposed them about Colesberg, moved to drive in the British flanks and force a retreat down the railway line towards Naauwpoort. Clements' left flank rested on a low rocky ridge known as Pink Hill, which was held by 75 men of the Victorian Mounted Rifles [including Tom Stock], 20 South Australians and 100 British regulars consisting of 50 Inniskilling Dragoons and an equal number of infantry of the 2nd Wiltshire Regiment.

When the enemy assault began shortly before noon, carried out by the bulk of General E.R. Grobler's force of 1,000–2,000 men, the post was commanded by Major G.A. Eddy[10] of the Victorian defenders, who had just taken over from an Imperial officer that morning. Soon the defenders were being lashed by close-range fire from three pom-pom quick-fire weapons and a field-gun, as well as innumerable Mauser rifles. Throughout the battle Eddy moved among his men, giving encouragement and directing their fire. After two hours, however, it was obvious that the position must eventually fall. Eddy accordingly instructed the Wiltshires to retire, while the mounted troops [including Tom Stock] continued to provide covering fire. Once the infantry was safely away, he then gave the order for the rest to follow, but no sooner had he done so, than he was killed by an enemy bullet through the head. The action ultimately cost six Australians killed, [Tom Stock among them] and one of whom died of wounds the following day, and 22 wounded; two of the wounded and one other man were taken prisoner, although one of these (a wounded officer) was released the following day. The gallant defence maintained on the hill ultimately proved to be a futile gesture. By 3.00 pm Pink Hill was in Boer hands. Since the right flank was also successfully pushed back by the Boers, Clements had no alternative, but to withdraw south and adopt a new defensive line around Arundel. It was a defeat which led to a major withdrawal from the area; it involved proportionately high casualties, but it was also praised as a great display of Australian courage and honour in war. Apart from the laudatory comments made at the time about the Australians' performance, however, there was also some criticism of Eddy for not having given the order to evacuate sooner and reduce needless losses.[11]

QUEEN VICTORIA'S CHOCOLATE

In 1900 Queen Victoria conceived the idea of sending a personal New Year's gift to British soldiers serving in the Boer War, a gift later extended to include colonial troops and members of the naval brigade. Three of the major British chocolate manufacturers, Fry's, Cadbury's and Rowntree's, produced the tins of chocolate. Initially, these manufacturers were reluctant to support the war effort, because of their Quaker pacifist ethic, but were persuaded to change their minds, although they had decided that the tins would carry no brand name. Queen Victoria was not amused by this decision, as she wanted her troops to know she was sending them quality British chocolate. As a compromise the Cadbury name appeared on the interior packaging of the chocolate. Fry's designed the tin which was copied with slight variations in design, colour and size by the other two firms. The small tins had a gold or blue coloured rim around the edge of the lid and in the middle was a gold-coloured embossed picture of Queen Victoria' head, her personal insignia, the words South Africa 1900, and 'I wish you a Happy New Year.' It also carried her signature. In all, 120,000 tins were distributed in South Africa which went on well into the New Year. Queen Victoria paid for the gifts from the Privy Purse and the cost of getting them to the soldiers was donated by various railway companies and the Union Steamship Company. Many men treasured their personal gifts from the Queen – some claimed they 'were as good as a medal.' Such was the prestige of the gift that a considerable number were immediately posted home unopened, with wrappings marked 'With Care, Queen's Chocolate'. **Company Sergeant-Major Ernest Coffey** mailed his tin of chocolate home to his church minister in Hawthorn. The empty boxes, often carefully preserved for generations, were commonly used to hold the service medals issued to Boer War veterans. Sources – (AWMREL/02572) https://www.angloboerwar.com/forum/19-ephemera/10607-siege-of-kimberley-framed-chocolate-tin

Sergeant-Major Coffey also related the story of this action from the point of view of the nearby VMI, in his letter home in which he described being a participant in the action, but at a distance.

On 10th February a half company of ours moved out to reinforce the V.M.R. and two squadrons of the Inniskillings at Windmill. Their picquets had been driven in, and one sergeant and one private of the V.M.R. killed. Our remaining half company occupied a kopje west of Kloof. An artillery duel from both sides was going on for about three hours on our left front. Our orders were to wait for the enemy to make a break and then cut them off. I predicted to Lieutenant Pendelbury, in command of our half company, that there would be no break of the enemy, as reinforcements were being rapidly pushed out from Colesburg. We passed a watchful night on our post, as the enemy were getting bolder and appearing in large numbers all round our front and flanks. We got no news from the Windmill that night, and were anxious, as heavy rifle fire was going on there, and the artillery had to withdraw at a gallop under a heavy fire of shrapnel from a 40-pounder battery. In the morning about daybreak, word came to us to stand by to retire from our position, as a result of the fighting on our left. Towards midday, after severe fighting, our men had to beat a hasty retreat, leaving Major Eddy, Lieutenant Powell, and Corporal Ross dead. [also, Private Tom Stock][12]

Other Victorians were killed and wounded in this action and Coffey wrote that it was 'with sad hearts and the enemy almost around us,' that they had to vacate their position and begin a retreat to Rensburg. At 10.00pm that night they halted and fearing the Boers would be in pursuit, built defensive stone positions and rested for a few hours. Eventually they met up with the rest of their company and the remainder of the VMR and, with both men and horses mad for water, reached Rensburg at 6.30am. The British forces were in full retreat and the men passed many bonfires of stores, fodder and vehicles on their way to their destination of Arundel. The Australians were soon joined by two British infantry regiments, the 2nd Battalion of the Wiltshire Regiment and the 2nd Battalion of the Bedfordshire Regiment, and in due course the pursuing Boer forces also appeared. The ensuing action was described by Coffey.

Major George Albert Eddy of Castlemaine (AWM PO4321.002)

The morning we arrived we were turned out to assist some companies of the

Wiltshires and Bedfords in getting into [Arundel] camp. Our retreat was not carried out too soon, as the enemy had pushed on rapidly, and when we turned out, had to face a 40-pounder, [artillery piece] which came among us, too often and close to be pleasant. One company of the Bedfords came in, but two companies of the Wiltshires were cut off and captured after a loss of 36 killed and many wounded. These companies were on picquet [duty] at Rensburg and were not told when to retreat (this looks like an error on the part of someone in authority), and of course the column moved away without them.[13]

The enemy would get some good hauls [of stores] from what was left behind, one stack of fodder was estimated to be worth £11,000. We are all fairly knocked up, men and horses, but keep at it and wish for it soon to be over.[14]

The exact circumstances of Tom Stock's death were later detailed by his brother Duncan in an Australian newspaper article headlined, 'The Late Private Stock. How He Fell' in late March 1900.

An interesting letter has been received from Private Duncan Stock, who was with his brother Tom when he was shot at Colesburg. He says about the engagement:

In getting away over the kopjes (small hills), my saddle worked back, and I got a spill, but I managed to get my horse right again and put my saddle right under a hot fire. Just then one of our men named Boulding came up, so I got him on my horse and galloped into camp under a hot fire of rifle and shell without getting touched. Then I was put into the trenches at camp, but Tom had to go on picquet duty at Hobkirk's farm at the foot of Pink Hill. We only had 160 men to fight 3000 or 4000 Boers. The fight lasted from daylight until 3.00pm. The Boers wounded most of our horse-holders and when our men had to retire, the fire was so hot on the horse, they thought it better to let them go and foot it back to camp. They had a hard fight and only nine out of eighteen came back. They were either wounded, killed or taken prisoners. Tom was doing good work, while making a hard fight to get away from the advancing Boers when poor Tom was shot in the head and dropped dead. The ambulance men got to him alright. Harry Bush, Tom's horse-holder got his thigh and wrists broken with a bullet. The ambulance men got him as he was wounded too badly to be of any use to the Boers as a prisoner. I don't feel in good spirits to write as I feel it (Tom's death) very much. I know you will be very much cut up about it.[15]

At home in the Casterton area, confusion spread as there was uncertainty as to which of the Stock brothers had been killed, and initially it was announced in the newspapers that Duncan was the victim. This news was conveyed to Duncan's distraught wife and children and other family and friends, until eventually a telegram was received from the Australian authorities confirming Tom's death; ironically on the same day that his family received his last ever letter. It was not until the Boers withdrew from their positions on Pink Hill in early March, that the bodies of the seven soldiers killed in the battle were able to be identified and buried where they lay. At the conclusion of the war in 1902, all the bodies were reinterred in the Memorial Garden of Remembrance cemetery of the nearby major town Colesburg.

In Australia family and friends soon gathered at Sandford on Sunday 11 March 1900 to pay their respects.

A memorial service for the late Tom Stock, who died in battle in South Africa was held at St Mary's Anglican Church. The Rev. J. Davidson preached an appropriate sermon and referred to the sterling qualities of the soldier who had laid down his life for the Empire. Mr Rogers of Henty, assisted at the service, by reading psalms and prayers and Lieutenant Little with several members of H Company VMR, occupied front seats in the church. The congregation was one of the largest that has been in the walls of the church. A wreath of everlasting tied with red, white and blue-ribbon streamers was placed on the cross on the altar and the church was draped in mourning. Special hymns were sung, and organist Miss Nellie Anderson played the Dead March from Handel's *Saul*, the congregation standing whilst it was played. The Rev Mr Davidson said that he had been made an offer from a person of a donation towards having a stained-glass memorial window put in their church at Sandford. He suggested that a committee should be appointed to take the matter up and collect subscriptions towards a fund to have it carried through, as a memento of the Hero and the words that would be engraved on it would a living remembrance of him.[16]

Tom Stock was the first Caulfield Grammarian soldier to serve overseas and sadly to lose his life in action.

Sir Arthur Conan Doyle, the author of the Sherlock Holmes books, served as a medical doctor in South Africa during part of the Boer War. He later wrote in glowing terms of the engagement at Pink Hill and the involvement of the Victorians, although the numbers he quotes were not always quite accurate.

The gallant Australians [Victorian Mounted Rifles] lost Major Eddy and six officers out of seven, with a large proportion of their men; but they proved once and for all that amid all the scattered nations who came from the same home, there is not one with a more fiery courage and a higher sense of martial duty than

THE VFL AND THE 1898 GRAND FINAL

At the end of the 1896 Australian Rules football season in Melbourne, eight leading clubs Carlton, Collingwood, Essendon, Fitzroy, Geelong, Melbourne, South Melbourne and St Kilda, broke away from the Victorian Football Association to form the Victorian Football League (VFL). Some innovations for the new competition saw the introduction of a finals tournament and the formal establishment of the modern scoring system with six points for a goal and one point for a behind. In 1897, Essendon became the Premier team, decided by the results of a round robin system and no Grand Final was played. The 1898 season comprised 14 rounds with the eight clubs playing each other twice and the teams fielded twenty players a side with no reserves. The first ever VFL Grand Final was played between Fitzroy and Essendon in appalling ground surface conditions at the Junction Oval in front of a crowd of 16,538 people. Fitzroy took an early lead, and Essendon was never able to catch up. **Stanley Reid** played a very strong game on the full-back line and was named amongst Fitzroy's best players. An Essendon opponent in the game was their full-forward **Charlie Moore** who kicked one goal in a losing score to Fitzroy. He later enlisted in the 4VIB and died of wounds on 12 May 1901, just five weeks before Reid was wounded in action and later died as a result.

the men from the great island continent. Be it said, however, once and for all, that throughout the whole [British] African army, there was nothing but the utmost admiration for the dash and spirit of the hard riding, straight-shooting sons of Australia and New Zealand. In a host which held many brave men, there were none braver than they.[17]

Two weeks after the death of Tom Stock, Lt. George Macartney was very fortunate not to suffer the same fate.

In the final assault on the Boer position on 15 December General Barton's task was to take Pieter's Hill. His force that day was the Royal Irish, Royal Scots, and Royal Fusiliers [Macartney], and right well did they carry out their mission. Had that assault miscarried, the remainder of the operations would have been at a standstill. Of the fourteen days' fighting, the battalion's losses were 1 officer, and 3 men killed, and 4 officers [including Macartney] and 70 men wounded.[18]

The CGS Speech Night programme of 1900 reported that:

Lieutenant [George] Macartney, son of Rev. H. B. Macartney, who was with the 2nd Royal Fusiliers and who was dangerously wounded in the storming of Pieter's Hill, during the relief of the city of Ladysmith, was also many years at Caulfield Grammar School.[19]

What the programme did not reveal was the seriousness of his wounds and this was reported some years later in an Australian newspaper.

Lieutenant (now Captain) Macartney took part in the Boer War where he was severely wounded in the storming of Pieter's Hill by a Mauser bullet which passed through his head [travelling from ear to ear]. For three weeks he was unconscious and at the end of that time a successful operation was performed, but he remained paralysed for a considerable period. His memory, however, was greatly impaired and his education had to begin anew. While studying, he travelled extensively and having regained most of his lost knowledge again, entered for his military examinations, which he passed successfully. The strain involved by his studies having proved very great, Captain Macartney [resigned his commission and] went to Canada where he took up farming.[20]

In late February another Caulfield Grammarian sailed for South Africa in the person of Stanley Reid, who had only spent a short time at CGS before transferring to Scotch College. He had a glittering sporting career and during his time at Scotch College he played in the 1st XI, taking 11 for 55 in one match against Melbourne Grammar School. Stan was a member of the Scotch College 1st XVIII and after leaving school, played 24 games for the VFL Fitzroy Football Club between 1897 and 1898 and played in the winning 1898 VFL Grand Final. Poignantly one of his Essendon opponents that day was Charles Moore, who as a member of the 4VIB

Stanley Spencer Reid in his final year at Scotch College (https://commons.wikimedia.org/wiki/File:Stanley_Spencer_Reid_at_Scotch_College.jpeg)

in South Africa, died of wounds after a gun battle on 12 May 1901. The CGS
Speech Night program for 1891 noted that in the Melbourne University Ordinary
Exams that former old boys, W. Boyd and S.S. Reid had both passed First Year in
Arts. He received his B.A. at Melbourne University in 1896 and then trained as a
theologian at Ormond College until his ordination.

In 1898 he took up the position as Minister at the Boulder City Presbyterian
Church in Western Australia, but when the opportunity arose to enlist in the 2nd
Contingent of the West Australian Mounted Infantry (2WAMI), Stanley took
leave of absence from his parish and as Private S.S. Reid (no. 41), embarked from
Fremantle with the rest of the eight officers, ninety seven men and 125 horses on
3 February 1900.

The 2WAMI arrived at Cape Town on 24 February and during their time of
service, according to the Official Records, took part in dealing with 'organised
disaffection' amongst the civilian population. They arrived at Blomfontein in
time to be attached to British General Pole-Carew's 11th Division, where they
served until October and joined in the advance to Pretoria. Amongst other places,
they saw action at Johannesburg and Diamond Hill in mid-June 1900. They
were tasked with patrolling the railway into the Orange Free State and eventually
reached Middelburg, 90 miles east of Pretoria on 27 July, having been involved
in actions at Vet River, Johannesburg, Silverton and Diamond Hill. Lord Roberts
in his telegram of 16 June said:

> [Louis] Botha's army has retired, believed to be in Middelburg. His rear guard was
> surprised and thoroughly routed by Ian Hamilton's Mounted Infantry, chiefly the
> West Australians and the 6th Battalion.[21]

Eventually, in late July, the 2WAMI took part in the advance from Pretoria to
Komati Poort and took part in the fighting around the Belfast District. In late
September they were also successful in recovering a large amount of captured
railway material and stores. Stan Reid and another 2WAMI soldier, Trooper
John Campbell, through a chain of unforeseen circumstances, became separated
from the main unit and in late July and early August were Reported as Missing
for two weeks. Eventually re-joining the 2WAMI, they, along with many of the
unit, were given the chance to return home, an opportunity that Reid took up.
Unfortunately, a personal letter he had written home, in which he had been
most scathing about one of the WAMI officers, had been published in a WA
newspaper. Consequently, Stan Reid was arrested in order to be put on trial on
his return to Australia and was even held under arrest on his voyage back home.

As events transpired, it turned out that no charges were eventually laid against him and in fact, the military authorities offered him a lieutenant's commission to help to clear his name!

The 2WAMI ended their days in South Africa on 7 November 1900 when they sailed for Fremantle landing on 8 December, having suffered no battlefield casualties.

While one Grammarian returned home, others were on their way to join the fight.

Endnotes

1 Droogleever. R. *Colonel Tom's Boys. Being the Regimental history of the 1st and 2nd Victorian Contingents in the Boer War.* 2013. PrintBooks, South Melbourne. p88

2 Wilmot. P. (Ed) Tom Stock Letters published in the *Casterton News.* Quoted from *The Second Harvest – The Writings of Tom and Duncan Stock, two Victorian brothers in the Boer War, 1899–1902.* 2000. pp16–17

3 Wilmot. P. ibid. pp15–16

4 Wilmot. P. ibid. pp16–17

5 *Age* Friday 23 March 1900. p6

6 The 37 mm Vickers Nordenfelt and or Maxim [Pom-pom] gun fired a one-pound shell at a rate of 60 rounds per minute and had a range of 2700 metres. Originally rejected by the British Army, the Transvaal Republic quickly adopted them as a cost-effective way of getting a light support weapon into the hands of each of their commando forces. The shells had smokeless powder, and sudden bursts could cause chaos as unsuspecting men and horses fell or bolted for cover. The Boers became adept at 'shoot and scoot' missions by firing 30 to 50 rounds and then quickly changing position. Given the sound they made with one round per second, the gun became known as the 'pom-pom.' Eventually they were also issued to the British Army, and very good use was made of them by Australian forces.

7 *Age* Friday 23 March 1900. p6

8 Wilmot. P. ibid. p17

9 Wilmot. P. ibid. p17

10 Major George Albert Eddy was a former schoolteacher who became a professional soldier and was made second in command of the Australian Regiment in South Africa. The night before this action he 'stood on the very spot where he was to die the next day and remarked to friends that the position was superb for a last stand.' For his bravery in the action where he was killed, he was Mentioned in Despatches. (AWM Last Post Ceremony 11 July 2013).

11 Coulthard-Clark. C. *The Encyclopedia of Australia's Battles.* 2001. Allen and Unwin. Crows Nest. NSW. p66

12 *Age* Friday 23 March 1900. p6

13 The Wiltshire's Commanding Officer, General Clements found it necessary to order a retirement from the Rensburg positions on Arundel at 5.00am but later altered the time to 12.15am. The notice of the alteration was not sent to the two companies on the outpost and when they returned to camp, found it had been occupied by the enemy. Trying to make their way to re-join the rest of the force, the Wiltshires were surrounded, and most of them taken prisoner. (https://www.angloboerwar.com/unit-information/imperial-units/2052-wiltshire-regiment)

14 *Age* Friday 23 March 1900. p6

15 *Australasian.* Saturday March 24, 1900. pp638–9

16 Wilmot. P. ibid. p18

17 Doyle. A.C. *The Great Boer War.* Nelson and Sons. London. 1903. p188

18 https://www.angloboerwar.com/unit information/imperial units

19 Caulfield Grammar School Speech Night program. 1900. p10

20 *The Farmer and Settler.* (Sydney. NSW). 6th July 1915. p2

21 Murray. P.L. *Official Records of the Australian Military Contingents to the War in South Africa.* 1911. pp403–404.

CHAPTER 5

THE BOER WAR – CONVENTIONAL WAR PHASE 2 CONTINUES

Late February 1900 saw the arrival in South Africa of Walter Laishley Spier and Frank Weir, two Caulfield Grammarians who had enlisted in the New South Wales Citizen's Bushmen Contingent (NSWCBC). This unit was raised in the first instance by public subscription; hence the name. As outlined in the recruiting notices, the object was:

> … to enrol a regiment of countrymen acquainted with the vicissitudes of bush life, good shots, good riders and of sound physique – such a class of men, in fact as would be fitted to cope with the enemy, according to the methods of the latter. Preference, therefore, would be given to men who had previously served in South Africa and/or those having experience of country work in Australia, management of horse and bush travelling.

The Australian War Memorial noted that, the New South Wales Citizens' Bushmen was the third contingent sent by New South Wales to the war in South

Frank Weir (https://www.caulfieldgrammarians.com.au/alumni-profiles/frank-weir)

Walter Laishley Spier (https://www.caulfieldgrammarians.com.au/alumni-profiles/walter-laishley-spier/)

Africa and consisted of 30 officers and 495 other ranks, with 570 horses.

Walter Spier, a former CGS boarder, had moved to Sydney and after leaving school had become a station overseer. He enlisted on 13 February 1900 as a Corporal (152) in B Squadron of the New South Wales Citizens' Bushmen and served from April 1900 – April 1901 in Rhodesia. Following this he served in the West Transvaal and in northern Transvaal.

Frank Weir undertook two separate periods of service in South Africa with his first period also being as a member of the NSWCBC. With his rural background, and ideally suited for this work, Weir enlisted (no. 15) in A Squadron of this unit and was listed in its original roll as being aged 22 years and 10 months old, 5' 8½" in height, and a station overseer of North Yanco, near Leeton in NSW.

BATTLE OF ELANDS RIVER

This battle was fought between 4–16 August 1900 and involved the heroic defence of a staging post in the Western Transvaal. **Frank Weir** and **Walter Spier** were members of the mixed force of 300 Bushmen from all the Australian colonies except South Australia. The forces were also augmented by another 200 troops from Rhodesia, Canada and three from British units. Apart from their rifles, the small force only possessed two old field guns. The Boers had amassed some 2–3000 men armed with at least five 12m pounder field guns, 3 pom-poms and a Maxim gun. The task of the British garrison was to guard a substantial store of supplies intended for other British columns, hence there was an intense interest in its capture by the Boers. Situated on a small rocky ridge in the middle of a natural bowl, small detachments of British troops held other small hills and defensive positions on the riverbank. During the first two days of the attack the Boers hit the camp with at least 2,500 shells which killed most of the horses, mules and oxen. Relief columns were unable to come to the rescue and eventually the siege turned into a bitter stalemate. The Boers did not want to destroy the camp as they desperately wanted the stockpiled British supplies. Consequently, accurate small weapons fire from the Boers saw casualties mount and the British troops eventually were pinned down during daylight hours. Most of the wounds suffered by the troops were inflicted by artillery shells. Eventually, a large British force of 10,000 troops relieved the garrison after twelve days of siege. Jan Smuts the Boer commander praised his enemy, *'All honour to those heroes who in the hour of trial rose nobly to the occasion.'*

The unit left Sydney on 28 February 1900, on board the transports *Atlantian* and *Maplemore*, and arrived in Cape Town on 2 April. The ships then proceeded to Beria, where they disembarked on 12 April and headed to Bulawayo, Rhodesia (now Zimbabwe).

The regiment was divided into four mounted rifle squadrons and the regiment's staff, along with A Squadron (Weir), served under General Plumer, participating in the relief of Mafeking in May. The squadrons were dispersed into various areas and served in Rhodesia, north and west Transvaal in the hunt for the Boer leader Christiaan de Wet, and the advance on Petersburg. At one stage the Bushmen were left behind as garrison troops to carry out patrolling, scouting and other general tasks. Weir commented in his diary that they had been left behind:

> … owing to our Regiment being so small and our officers not in favour. Our work is added to by the Division leaving 56 of their sick horses to attend to and the work is so heavy that even as a Sergeant-Major, I must take a patrol myself in the morning.[1]

Still under Plumer's command, A Squadron subsequently also served as part of General Baden-Powell's column in the relief of Rustenburg in July. The rest of the regiment, B (Spier), C, and D Squadrons took part in a heavy engagement at Koster's River on 22 July 1900, to relieve the Eland's River garrison. B Squadron (Spier) was engaged in heavy fighting at Megato Pass. A Squadron (Weir) was part of a force which defended the Eland's River post against a Boer force of about 1,000 men for 13 days until they were eventually relieved by Lord Kitchener's force on 15 August. Lord Methuen's relief column arrived three days later.

Two more Grammarians sailed from Melbourne on 28 February 1900 with the contingent of the Third Victorian Imperial Bushmen (3VB) contingent. The unit consisted of about 250 soldiers that were kept together in South Africa, as there were only two squadrons, and for most of that time they served under the same commander.[2]

James William Christie transferred from CGS to Melbourne Grammar School and upon leaving there commenced work as a surveyor's assistant with the Melbourne Metropolitan Board of Works. He had enlisted as a Private (491) in B Squadron of the 3VB. Arnold Davies, a Malvern Grammarian also enlisted as a Private, but in A Squadron. He was the son of Sir Matthew Henry Davies who built Valentines Mansions, now the Malvern Campus of Caufield Grammar School.

The unit disembarked at Beria in Portuguese East Africa (Mozambique) on 11 April 1900 where, along with other units, all the colonial Bushmen were formed

James Christie *(Droogleever)*[3]   Arnold Davies *(Droogleever)*[4]

into regiments known as the Rhodesian Field Force (RFF). Trains took them on part of their journey, but they had to march in squadrons through Rhodesia to Bulawayo and from there to Mafeking. Here they were again mobilised, equipped and moved out in regiments under the command of Lt. General Robert Baden-Powell,[5] with their destination being Rustenburg.[6]

The Third Australian Bushmen regiment, two thirds of which were the 3VIB, became a vital core of Baden-Powell's force in keeping communications open between Mafeking and Pretoria after the relief of Mafeking. The 3VIB and some of its squadrons were involved in actions or battles at Koster River, Elands River, Rhenoster Kop and Pietersburg. Arnold Davies was a key participant in 3VIB's first action at Koster River on 22 July when seven members of the unit were killed. A letter he wrote home about this action was published in a Melbourne newspaper and he made some pertinent observations.

> I suppose you have heard by now that we had a go with the Boers last Sunday and the wonder is that any of us came through alive. At about 7.00am we rode thinking of when we could get breakfast, when suddenly they opened fire on us from two hills on the left of the road. We had ridden fairly into a trap. We dismounted but were lying out in the open and if a man made the slightest movement, he brought down the whole of the fire on himself. We stayed there until about 1 o'clock when we made a dash for a drain some 30 yards away, but two of our men were killed while they were running. Once in the drain things were more comfortable. But

BRITISH MEDICAL AID

The British Army had only established the Royal Army Medical Corps (RAMC) in 1898. It had never served in a major conflict and lacked any experience and efficiency. In time, the RAMC did prove to be very successful in treating over 22,000 soldiers for wounds and injuries but failed in the treatment of twenty times that number for preventable diseases. The British Red Cross was not formed until 1899 and was largely disregarded by the Army. Despite this, the Boer War proved to be a benchmark in the way that armies took care of their sick and wounded soldiers in times of war. For example, the discovery of X-rays in 1895 had greatly assisted physicians and surgeons in their work in the Boer War. In addition, much knowledge was gained about bullet wounds caused by the new high velocity bullets which had replaced the soft lead bullets of previous wars. The British RAMC's hospital arrangements in South Africa were generally divided into three types. Mobile Field Hospitals had ground sheets only, with no beds or stretchers and space for about 100 men. Stationary Hospitals were situated along main roads or railways which could treat about 100 men on stretchers. General Hospitals were usually situated at main bases and were as fully equipped as civilian hospitals and could deal with between 250 and 500 patients. There were also private hospitals which had been endowed by wealthy patrons. In addition, there were four hospital trains and ten hospital ships, some of the latter also privately endowed. Organised independent military nursing services also began, and several Australian colonies sent contingents.

we could see nothing to shoot at. I was lying on my back, when a bullet hit the ground at the back of my head and threw dirt over my face. That was the closest shave I had.[7]

When Baden-Powell left to raise men for the South African Constabulary, the Australian Bushmen continued their same valuable work under General Henry Plumer.

In common with similar formations raised at this time, such as the NSWCBC, the New South Wales Imperial Bushmen (NSWIB) was raised in response to a request from the British Government and what was asked for were 'hardy bushmen' who could ride, shoot and navigate in the bush. The idea was to find men who could employ the same type of guerilla tactics as the Boer and beat them at their own game.[8]

Caulfield Grammarian, James William Campbell served with them during their time from April 1900 to May 1901. With a strength of forty officers and over 700 mounted troops, the unit was renamed the 6th Imperial Bushmen and first served with the Rhodesian Field Force. Later it saw action in the Western Transvaal and was involved in the capture of the Boer General Koos de la Rey's convoy and guns. James Campbell survived the war and returned to Australia with the unit in July 1901.

May 1900 saw the largest influx of Grammarians in the same unit to arrive in South Africa, when six former Caulfield men arrived as members of the Fourth Victorian Imperial Bushmen (4VIB) contingent spread across five different squadrons, A, B, C, D, E. They were Andrew Anderson and Arthur Williams (A), Edward Duncan (B), Reginald Holloway (C), Thomas Foster (D) and William Boyd (E). The five squadrons consisted of nearly 630 men accompanied by nearly 800 horses, 11 wagons and other associated equipment. After parading through the streets of Melbourne on horseback, the Contingent sailed on 1 May 1900, reaching Beria in Portuguese East Africa 22 days later.

A musician by trade and unmarried, Andrew Anderson enlisted in A squadron as a Private with the regimental number of 333. He was 5' 8" with a chest measurement of 38 inches and was promoted to a Lance Corporal. Arthur George Thomas Williams enlisted as a private no. 213, but was reported to have been invalided back to Australia, arriving on 8 November 1900. However, later he was reported fit and rejoined his unit on 23 January 1901. William Boyd enlisted as part of F Company but was eventually invalided home, arriving back in Australia in early October 1900. Edward Duncan, for reasons unknown, appears to have enlisted as a Private (493) in B Squadron under the name of his older brother James Thomas Duncan. Standing at 5' 9¾" with a heavy sandy moustache and thick set, Duncan, along with the two Wallace brothers, just before their departure wrote to the local newspaper in Bright and Wandiligong to express their feelings on being in South Africa.

CONTRIBUTIONS TO THE WAR. – In April 1900 three boys wrote to the people of Bright and Wandiligong, via the Alpine Observer, thanking them for the field glasses which they had been presented with on their departure for the Boer War in South Africa. At the same time, they pledged to do their duty for their Queen and country in the Transvaal. The three were Andrew and Alexander Wallace and Edward Duncan.[9]

Reg Holloway (Droogleever)[11]

Thomas Foster (https://dustyheaps.
blogspot.com/2017/07/more-lost-
australians.html)

A miner from Inglewood, Thomas Barham Foster enlisted as Private (367) in the 4VIB being assigned to D Squadron. Reginald James Holloway, standing at 5 feet 9 inches, enlisted as a private soldier, no. 268 and was listed as a single bushman from Tyntynder. He became a member of C Squadron.

The British Government had learned some important lessons from earlier conflicts with the Boers and now changed tactics and methods to better undertake a non-conventional type of warfare.[10]

The Official Record stated that:

Each company or squadron consisted of about one hundred and twenty officers and men and upon their arrival in South Africa, both 'A' and 'B' Squadrons were sent to Bulawayo and then to Mafeking, where they formed part of Brigadier-General Lord Erroll's Brigade. The three remaining Squadrons, 'C', 'D', and 'E' under Major Clarke, remained in Rhodesia at Marandellas, Fort Charter, Fort Victoria, Tuli and Bulawayo; being engaged on the lines of communication, until the end of the year, when they were ordered to Cape Colony.[12]

Landing at Beria in Portuguese East Africa (Mozambique) and along with other colonial units, 4VIB was formed into the Rhodesian Field Force. Medical facilities were very basic, and many soldiers caught malaria, which then manifested itself into fever once they arrived in cooler climes and several deaths occurred at Umtali.

A Squadron (Anderson and Williams) and B Squadron (Duncan) were sent to Bulawayo in Rhodesia and then by rail to Mafeking in the Cape Colony. In a short time, they relocated to Ottoshoop where they formed a section of the Brigade under the command of Lt. Gen. Sir Frederick Carrington.

Meanwhile, C Squadron (Holloway), D Squadron (Foster) and E Squadron (Boyd) remained under the command of Major L.F. Clarke on garrison duty in Rhodesia at various locations, mainly being engaged in protecting and servicing the lines of communication. D Squadron (Foster) had been stationed in Marandellas since early July and was mainly occupied with garrison duties, field work and drill.

A line of mounted troops of the 4VIB marching from Umtali to Marandellas. (AWM P0122.008)

Sadly, this unhealthy environment was where the first 4VIB death occurred when after just three months in South Africa, Thomas Foster, aged 25, died from enteric fever and dysentery at Marandellas Hospital on 22 August 1900. A New Zealand correspondent had written an article in July 1900 about a fellow countryman, Trooper Fred Saxon, who was suffering from fever in the Marandellas Hospital. The correspondent reported on the appalling conditions that existed at the hospital.

On my return to camp after an absence of two days, I found the place deserted. Camp had been 'struck,' and the squadron to which poor Saxon belonged had entrained for Marandellas. Knowing that a number of New Zealanders were in the hospital, an old building which had been re-erected at the end of the paddock a day or two previously, stretched out on a heap of straw in a more or less neglected state, I immediately proceeded to the hospital to see what had become of the unfortunate patients for whom I had many times felt inclined to intercede, but of course would only have laid myself open to be told to mind my own business, as it was no part of my business as a correspondent to interfere with camp arrangements. My feelings on entering the building may be more easily imagined than described, when to my surprise I found one solitary occupant, the emaciated form of Trooper Fred Saxon, once a fine, strong young fellow, whose acquaintance I had made during the voyage. There, on the dirty floor, amid swarms of flies, fragments of some decomposed food and dirty utensils, the poor fellow lay as helpless as a child, with nothing but this filthy chaos and the dirty, dismal walls of the building and complete solitude to cheer his rapidly expiring spirits. A second glance was not necessary to convince me that my appearance had intensified the cause of the poor man's grief, for his eyes were still full of tears, and no wonder. I never in my life saw such a scandalous state of things. If the poor man had been a common criminal, he could not have been treated in a more brutal manner. In reply to my inquiry as to his condition he said, 'I am well enough, only I feel dreadfully weak. I have been starved. I have had nothing to eat since evening, this was at 4.00 pm yesterday, when I had a bit of biscuit and a drop of tea. I am supposed to get beef tea and cornflour, but I have only had it once or twice. When I complained to the man who is supposed to attend to us, he told me to get up and attend to myself or go without. I have not been able to walk for some time, I have been too weak. No one seems to take any interest in us or care whether we live or die.[13]

It should be noted that at this time around 30 other men from 4VIB had been invalided home, mainly suffering from the effects of malarial fever and dysentery.[14] Although Foster was entitled to the QSA and one clasp for service in Rhodesia, it must have been of little comfort that he was too unwell to be invalided home to Australia for better treatment. Foster was buried in the Paradise Cemetery at Umtali in Rhodesia (Zimbabwe) and sadly with the passing of time, his metal headstone, along with those of many of his former comrades, has disappeared, meaning that he now lies in an unmarked grave.

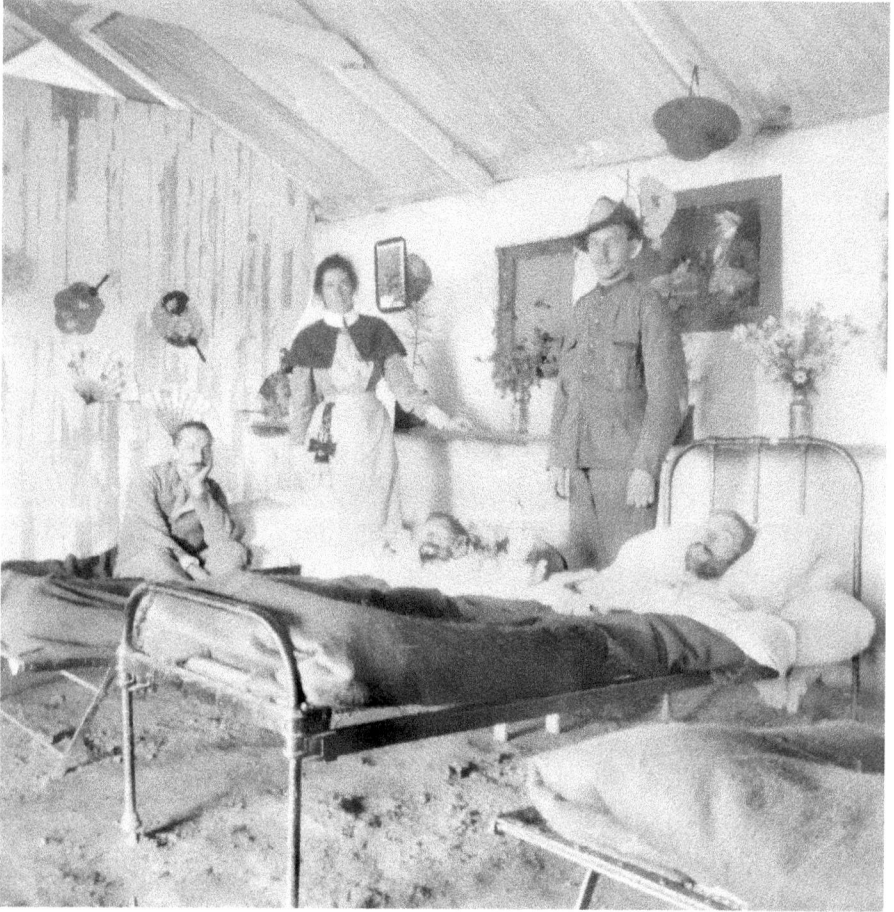

Informal group portrait of medical staff and patients in a hospital ward, possibly in Bulawayo, Southern Rhodesia [now Zimbabwe] circa 1900. The ward is decorated with flowers in a jar of water, pretty Japanese fans and framed portraits in an attempt to brighten up the primitive conditions as indicated by the bare earth floor. (AWM PO4544.011)

Ernest Coffey was still serving in South Africa and was named in a Melbourne newspaper in March 1900, when it was noted that the Rev. Herbert Taylor of Hawthorn received a box of the Queen's chocolate, a gift of Colour Sergeant E.N. Coffey, a former member of the Christ Church choir, who was with the First Victorian contingent. The newspaper wryly reported that the contents of the box when received, were still intact.

Ill health in the form of a serious chest infection, however, saw Ernest Coffey invalided to hospital in Capetown, and a newspaper report noted his military abilities in the field.

ENTERIC FEVER AT BLOMFONTEIN

Typhoid fever was known in Boer War times as enteric fever or bilious fever or Yellow Jack. It is a serious illness caused by the victim ingesting food or water contaminated with faeces from another typhoid infected person. Over 13,000 men died from typhoid, including Grammarians James Christie, Thomas Foster and Walter Spier. At least 31,000 British soldiers were invalided home suffering from its effects including Grammarian Bernard Bardwell, who was hospitalised in South Africa and later returned home to WA. Following the relief of Ladysmith on 1 March 1900, General Roberts took his army of over 33,000 men to Blomfontein, the captured capital of the Orange Free State for rest, resupply and reinforcement where they stayed until the end of April. Boer War historian, Thomas Pakenham wrote most disparagingly about the circumstances which led to the resultant typhoid epidemic. *'To prevent typhoid, you needed careful sanitation; to treat it you need careful nursing and a careful diet. Why was there such a typhoid epidemic at Blomfontein? Negligence was the very simple answer, neglect of simple sanitary precautions in the army camps and neglect of hospital patients.'* In addition, the Boer had captured the main water supply for the town, to add to conditions that led to over 1000 deaths in six weeks.

[CSM] E.N. Coffey showed himself a man for emergency and did grand service when Captain Pendlebury's company was almost denuded of its officers.[15]

Ernest Coffey was invalided to Australia in May 1900, arriving in Melbourne on 6 September 1900 and subsequently being awarded a pension of 3/6 per day. He was awarded the Queen's South Africa (QSA) Medal and was entitled to wear four clasps for the Unit's service in the Cape Colony, Orange Free State, Johannesburg and Diamond Hill. He had only been in Melbourne two days when he was welcomed home by his comrades of the 2nd Battalion Infantry Brigade at a dinner at the Queen's Café in William Street, Melbourne. Major T.J. Courtney, in the chair, said of Coffey that he had left Victoria under the command of the late Major Eddy, who was killed in action at Pink Hill. The Sergeant-Major had taken part in the march to Pretoria and so satisfactorily performed his duties that he had been promoted to the rank of regimental-sergeant major. A letter was read from Lt.Col. Burston the officer commanding the battalion, complimenting Coffey on his record and wishing him a speedy restoration to health. Major Courtney then proposed the toast to Coffey who spoke in a 'chatty and unaffected manner' in response, as reported in next day's newspaper article.

> He gave a brief outline of the movements of the First Victorian Contingent after landing in South Africa until the entry into Pretoria [in early June 1900]. His experiences naturally included many stirring incidents, all of which were told in the purely impersonal and modest spirit of one who prefers to speak of others rather than himself. It was while under the command of Colonel Tom Price that Coffey had his first experience of 'pom-pom' fire. It was near Brandfort, in the engagement which proved disastrous to Lieutenant Lilly and several other Victorians. The shells created a stampede among the horses and over a dozen were shot down in a moment or two. He [Coffey] was in the devil's own rage because he could not find out what the orders were [laughter] until Colonel Price gave the order to fall back but was himself the last to leave the position. He [Price] came strolling out of the firing line, just as leisurely as though he were smoking a cigar along Collins Street. At Kroonstad part of the corps were under a hail of hundredweights of bullets and projectiles from Mausers [rifles], pom-poms, Maxim-Nordenfelts and shrapnel – the result was one horse hit, and two helmets lost. [laughter] Under a similar variety of 'music' outside Pretoria there was one horse killed, and one man slightly wounded [laughter]. But it was beyond Pretoria that the doctor ordered him back to hospital and after a [very] slow recovery, he was invalided home.[16]

Coffey remarked that he would not go as far as to say that the Australians troops were better than the others there, but he was very sure they were as good as the best. He closed his response by paying tribute to the English soldiers, the 'Tommies', as wonderful men.

As a returned soldier from Richmond, he was amongst a small group who were also given a hearty welcome home by local dignitaries. A Melbourne newspaper reported under a bold heading that, 'Heroes at the Town Hall [who had] Returned from South Africa [were] Welcomed by the Council'.

> When the business of the Richmond Council was concluded last night (at an early hour), the mayor [Cr S.J. Willis] introduced five members of the First Contingent who had just returned from South Africa. Namely, Sergeant Major Coffey, Sergeant Dowd, and Privates Wood, Somerville, and Will who are all residents of Richmond. The mayor called for three cheers for them, and they were given with enthusiasm. An adjournment was then made to the mayor's room where they were suitably entertained, the mayor proposing their health in enthusiastic terms.[17]

The mayor's room had been decorated with flags and 'wore quite a patriotic appearance.' It was reported that the table had been well laid out with refreshments and when these had been partaken of, the mayor called upon the company to charge their glasses and toast the Queen. This was 'drunk right royally' with the company then all singing the National Anthem. The mayor then proposed the health of the returned warriors, and in the course of his remarks paid a high 'eulogium' to the bravery of the Australian troops. He also said that as all those [soldiers] belonging to Richmond returned, they would be welcomed by him on each occasion. The company then sang 'Soldiers of the Queen' and 'For they are Jolly Good Fellows' which was followed by a response from Sergeant-Major Coffey, who received quite an ovation on arising.

> He gave some very interesting incidents of the campaign and speaking of the high appreciation in which the Australians were held, instanced the fact that the Victorians had acted as advance scouts to Lord Roberts' army the whole route from Blomfontein to Pretoria. On behalf of himself and the other returned members of the contingent present, he returned the company his very best thanks for the honour they had done them.[18]

Some of the other returned soldiers also spoke and showed they had been through a hard time, not a bit of the kind of 'picnic' it was at first predicted they were going out to South Africa to take part in. Several other toasts were proposed, and a silent toast to those who had fallen on the battlefield brought the proceedings to a close.

The last Grammarian to join the conflict in 1900 was Malvern Grammarian A.W. McLean. A.W. McLean's details are few and far between insofar as official Malvern Grammar School records are concerned, and even his first two names are yet to be identified. In an early school magazine *Malvernian* Vol. 1. No. 4. 1907, McLean records that he was unable to enlist with the 4VIB regiment as he was only aged 19, so he sailed to Durban in South Africa and enlisted with the South African Light Horse, becoming its youngest Corporal. He also mentions that Bradley, perhaps a fellow Malvern Grammarian, also tried to enlist. An A. W. McLean is reported as being invalided home from Durban on the *Windsor* on 14 April 1901 and was visiting relatives in Australia.

The South African Light Horse (SALH) was raised in the Cape Colony as a volunteer mounted force in November 1899 and consisted of eight squadrons generally raised from the ranks of the [foreigners] 'Uitlanders.' They served at the siege of Ladysmith, the Tugela River conflict, and the battles of Spion Kop,

Vaal Krantz and Pieter's Hill. At this last engagement the SALH fought alongside George Macartney's Fusilier regiment, but at a time before A.W. McLean had joined them.

The latter months of 1900 saw much change in the state of play in the Boer War. Peter Dickens, a contemporary South African historian of the conflict, makes the following assertion:

> In terms of timing, the British victory in the Conventional Phase 2 is swift. From October 1899 – July 1900, a mere 10 months, the British had reversed an invasion, captured two separate countries, taken both Boer Republic's capital cities, taken the Boer's economic hub, isolated both countries and starved them of external aid. They had also broken the critical mass of the Boers to fight conventionally, captured nearly every major gun and artillery piece and occupied all the Boer's fortifications and defences. By any military standards that was good strategic, operational and tactical command.[19]

Given this situation on the ground in the contested areas of conflict, the Boer forces were left with the option of surrender or fighting on, but using different tactics and methods. Accordingly, the war moved to Phase 3 and became a conflict of the Boer employing guerrilla/insurgency tactics, consequently ensuring that the British moved to a war of counter-insurgency measures.

For the Caulfield and Malvern Grammarians involved in the conflict in the coming 20 months until the end of the war, there was much to be done in the service of the Empire, sometimes involving loss of further lives.

Endnotes

1   Weir. F.V. Diary held in Mitchell Library, Sydney.

2   Drooglever. R. *That Ragged Mob – Being the Service Record of the 3rd and 4th Victorian Bushmen Contingents in the Boer War*. Trojan Press. Melbourne. 2009. pxi.

3   Droogleever. R. *That Ragged Mob*. ibid. p515

4   Droogleever. R. *That Ragged Mob*. ibid. p515

5   Robert Baden-Powell was a Lieutenant-General in the British Army, and later the founder and Chief Scout of the world-wide Scout movement.

6   3rd Victorian Bushmen. Outline. https://alh-research.tripod.com/Light Horse/Index.blog/topic id-1115696

7   *Argus* Friday 14 September 1900. p5.

8   https://www.awm.gov.au/collection/U52012

9   Lloyd B and Nunn K. *Bright Gold – The story of the People and the Gold of Bright and Wandiligong*. Histec Publications. 1987

10   Murray, P.L. ibid. pp252–257.

11   Droogleever. R. *That Ragged Mob*. ibid. p597

12   Murray, P.L. ibid.

13   *Feilding Star* of 5 October 1900 ui

14   Droogleever R. *That Ragged Mob*. ibid. p264

15   *Herald* Wednesday 29 August 1900. p1. Captain Henry William Pendlebury commanded No.4 Company of the VMI from 7 April 1900 onwards and the unit was involved in numerous actions including being present at the surrender of Johannesburg and Pretoria.

16   *Argus*. Monday 10 September 1900. p5

17   *Argus*. Friday 7 December 1900.

18   *Richmond Guardian*. Saturday 8 December 1900. p2

19   https://samilhistory.com/2023/10/26/the-boer-war-myth-busting-by-the-numbers/

PART THREE

THE BOER WAR – GUERILLA WAR AND COUNTER-INSURGENCY

CHAPTER 6

PHASE 3 (PART 1) –
SLOW PROGRESS TOWARDS
BRITISH SUCCESS
SEPTEMBER 1900 – MAY 1902

The war now settled down into a long grind of guerrilla warfare by the Boers and British mounted troops that operated counter-insurgency measures. Many Grammarians participated in this phase, which involved very few large set-piece battles, instead consisting of skirmishes and raids of diverse sizes as well as interminable patrolling to harass and neutralise the Boer commandos. In time the British adopted a system of constructing blockhouses along supply lines and destroying Boer farms, livestock, and homesteads to eliminate sources of food, comfort, and support for the Boer. Eventually the British moved many Boer families into what was termed 'concentration camps' where many died because of poor conditions and disease. These tactics, in time, saw the British be victorious, but they also led to much condemnation from many quarters, especially when it was revealed that 27,000 Boer women and children had perished in these camps. In addition, many Australian soldiers came to hate the practice of burning houses, crops and homesteads and evicting women and children from their farms. Many of the soldiers came from rural backgrounds themselves, including numerous Grammarians, and they secretly empathised with the Boer families.

The new Australian Commonwealth government had been proclaimed on 1 January 1901 when the former colonial governments had embraced Federation. Draft contingents of mounted troops were raised by state governments after Federation on behalf of the new Commonwealth Government, as the Federal body was unable to do so at that time. Some of these contingents did not reach South Africa until March and April 1901. The Australian Commonwealth Horse (ACH) contingents consisting of eight battalions were eventually raised by the new Federal government and became their first ever expeditionary military force. The 1st and 2nd Battalions arrived in Durban in March 1902 and saw limited action until the end of May. The 3rd and 4th Battalions arrived in late April and, awaiting movement orders, went into camp where they stayed until the end of May. The majority of the third contingent consisting of the 5th, 6th, 7th and

AUSTRALIA BECOMES A FEDERATION IN 1901

There had been calls from the Australian colonies late in the nineteenth century, for the country to become a federation of states. Sir Henry Parkes, five times the Premier of NSW between 1872 and 1891 gave a speech at Tenterfield NSW in 1889 and called for a great national government for all Australians. Parkes also argued that federation would make it easier for each colony's military to unite as a single national army under the command of one single Federal Government. A major outcome of his speech was a call for an Australian Convention where the parliamentary representatives of each colony could debate the issues and work to develop a Federal Constitution. These Conventions took place in 1891 and 1897–8 and a Constitution Bill was drafted in 1898. Two rounds of referenda were needed, and a Yes majority was secured in all colonies except WA. The British Parliament approved the Bill on 5 July 1900 and Queen Victoria then gave the legislation royal assent and declared it would take effect on 1 January 1901. WA voted again on the issue on 31 July 1900 and a Yes majority was achieved. On 31 December 1900, the Australian Governor General, Lord Hopetoun swore in the first Australian federal ministry with Edmund Barton as the caretaker Prime Minister. In March 1901 the Australian Government assumed all responsibility for matters of defence and the six colonial armies were merged into the Commonwealth Military Forces. The first federal election was held on 29–30 March 1901 with Barton continuing as Australia's first Prime Minister and the first parliament opening in Melbourne on 19 May 1901. Australia had become a self-governing nation.

8th Battalions were still at sea enroute to South Africa when peace was declared.[1]

Eventually, in May 1902, the British and Boer forces agreed on a final peace treaty in which the Transvaal and the Orange Free State agreed to become British colonies in return for a pathway to eventual self-government.

Meanwhile, from September 1900 onwards, the war continued to grind on. On 4 December the 1VMR and VMI returned to Australia following their term of service, having lost seventeen members to the war including Caulfield Grammarian Tom Stock. Other units had also completed their tour of duty and on 8 December the 2WAMI, including Stanley Reid, returned to Fremantle, fortunately having suffered no battlefield casualties during their tour.

A Boer farm being burned. (https://samilhistory.com/2024/02/04/war-is-cruelty)

Arnold Davies and James Christie, members of 3VIB, were engaged in the unit's first action at Koster River on 22 July 1900 where seven members of the unit were killed in heavy fighting. Former Caulfield and Melbourne Grammarian James Christie had enlisted in the Third Victorian Bushmen and, since the unit had arrived in South Africa on 11 April 1900, had seen much action. In November 1900 with B Squadron, he was involved in the unpleasant duty of clearing Boer women and children from farms which were then put to the torch. Just a few weeks later he contracted enteric fever and was admitted to hospital at Rustenburg, where he died on 7 December 1900, the second Grammarian to succumb to this disease. His death notice in the newspaper read.

> Death. CHRISTIE – On 7th December 1900, at Rustenberg, South Africa, of enteric fever, Trooper James William Christie (3rd Bushmen's Contingent), only and beloved son of Catherine and the late William Christie, Grafton, and stepson of the late Captain John Muir, Manning River, aged 30 years.

Arnold Davies also fought at the battle of Rhenoster Kop on 29 November 1900 with a combined British and colonial force of over 2500 men. This proved to be the last pitched battle of the Boer War and cost the Boers 31 casualties and the British 85 casualties, including thirty among the mounted troops. After this exploit, however, the ranks of the 3VB were thinned and it was reported that.

> A Squadron (Davies) could only muster thirty men out of 125 as disease, exhaustion and exposure had decimated the ranks.[2]

By the time the 3VB's 12-month tour of duty expired in April 1901 and they returned to Australia in June, they had lost seventeen men. Eight were killed in action, two died from accidental injuries and seven succumbed to disease, mainly enteric fever, including Grammarian James Christie.

A, B and C Squadrons of the NSWCBC were involved in many small skirmishes and A Squadron (Weir) was also present at the battle of Rhenoster Kop. In 1901 the Bushmen continued to serve under Plumer, operating in the Transvaal and the advance on Pietersburg. It was in one of these actions when Weir was to figure prominently. On 12 January 1901, the Boer attacked the column, and Frank Weir was severely wounded; subsequently he was mentioned in despatches for his gallant actions under fire.

> On 12 January 1901 when the small convoy escorted by West Riding Infantry and Citizen's Bushmen detachments was about to be attacked, the wagons were halted and drawn up on the open veldt and the only cover was the long grass. The small force was dispersed as best it could be, and Sergeant-Major Weir took over command of the right front sector with a section of twelve men. The infantry soon began to run out of ammunition and a Corporal volunteered to run back to the wagons to get some more but was immediately wounded. Sgt-Major Weir dressed the Corporal's wound under fire but was himself shot in the face. Sgt Major Weir continued firing until he fainted from blood loss. The infantry had no ammunition left, men were being wounded and the Boer fire was coming from only about 150 yards away.

In a report about the action to the Chief Staff Officer on 18 January 1901, Lt. Colonel H.B Jeffries, Royal Horse Artillery mentioned:

> The distinguished conduct of Sergeant-Major F.V. Weir, 1st Australian Bushmen, who appears to have behaved with great gallantry. He was twice shot in the face and continued firing until he fainted.

A letter to Weir's mother at Deniliquin on 6 May 1901, from Sgt Charles Druitt, (No.14) a fellow member of A Squadron recalled the incident.

> He was hit through both cheek bones, causing the mouth to be slightly stiff and as an operation was necessary, he proceeded to England to Netley (Military) Hospital on 7th March 1901. He and I have been great chums all through the campaign.

Sergeant-Major Weir was mentioned in C-in-C's despatches on 8 May 1901. In the meantime, the regiment had embarked at Cape Town and returned to Australia on 9 June 1901, disembarking in Sydney two days later.[3]

Tragically, James Christie was not the last Grammarian to succumb to disease. On 23 January 1901, former Caulfield and Sydney Grammarian, Walter Laishley Spier died of enteric fever at Woodstock Hospital Capetown. At wars end it was estimated that around 8,000 British soldiers had died from typhoid. Away from these scenes, other Grammarians were participating in other actions. Being a member of C Squadron of 4VIB meant that Reg Holloway remained in Rhodesia from early June until the unit marched to Tuli, arriving there on the 18 October for garrison duty until 23 November.

Members of C Squadron 4th Victorian Imperial Bushmen, the unit of which Grammarian Reg Holloway and former VFL player Charles Moore were members. (AWM P01222003)

The Squadron then marched for two days to Pont Drift, again for garrison and patrol duty until 12 December, when they were ordered to return to Bulawayo. Holloway's 'C' Squadron was now joined by 'D' and 'E' Squadrons and the Official Record explained what now took place.

C,' 'D,' and 'E' Squadrons, under Major Clarke, were engaged at Matjesfontein, collecting stock, removing undesirables, etc, until early in February 1901, when they entrained for De Aar, (near Colesberg) being attached to Col. Hon.

BLOCKHOUSES AND CONCENTRATION CAMPS

To combat the Boer troops being sustained by their families and supporters, Generals Roberts and Kitchener devised two schemes directly aimed at further isolating them: Blockhouse Forts and 'tent-village' Concentration Camps. **Blockhouse Forts** originally were a string of simple forts reinforced with stone walls and surrounded by defensive ditches and barbed wire. Eventually the large cost of building each structure led to the manufacture of prefabricated forts. Kitchener expanded this system by dividing the country into small grid areas, with fortified lines which in turn prevented or hampered the Boer from moving freely around the countryside. By war's end over 8000 blockhouses had been built which in turn absorbed 50,000 men to guard them all. Separate **Concentration Camps** were established for the Boers and Africans because of the British Army's 'Scorched Earth' policy of destroying Boer crops, livestock, homes and farms to deny their enemy a means of re-supply. Tens of thousands of Boer women and children were housed in the British Army administered camps. Soon, badly overcrowded conditions arose with very poor hygiene and sanitation, lack of adequate water supplies, meagre food rations, inadequate shelter and a lack of medical care. It is estimated that about 28,000 Boers and 20,000 Africans out of at least 100,000 inmates died in the concentration camps.

A.H. Hennikar's column. On 11th February, 40 men of 'C' Squadron, under Captain Tivey, made a forced march of forty miles to Philipstown and surprised the enemy, who numbered upwards of three hundred. They occupied the adjacent kopjes and who were driven back and kept in check until reinforcements arrived, when they retired.[4]

The Boer General Christian De Wet had re-entered the Cape Colony on 10 February and the focus of the British forces now turned to trying to capture him and his commando troops. Patrols were sent out to try to locate De Wet and also to contact other British patrols and it was on one of these forays that Reg Holloway was dangerously wounded in the head.

Lt Herbert Embling of 4VIB on 15 February 1901 near a farm named Wolverhinton, took just six men, Holloway amongst them, to search for the Boer leader. Unwittingly, Embling led them into a Boer ambush and three soldiers were wounded, two horses were killed and only one soldier managed to escape to sound the alarm. The remaining two men held off the Boers until the weight of numbers saw them captured and the Boers relieved the Australians of all their equipment and weapons. Holloway's condition at this point was reported by one of the two unwounded men.

> They [Boer] let me go about an hour afterwards and I fixed up the wounded. The poor chap that was shot in the head (Holloway) was paralysed and the officer (Embling) and me had to carry him four miles till we struck a farmhouse where we stopped until daylight and they got an ambulance from Colonel Plumer's column.

Charles Moore, the former Essendon footballer who had played against Stanley Reid in the 1898 VFL Grand Final, was also a member of this 4VIB C Squadron detachment and added his description of what had happened.

> You would not believe how the lead was flying. They [Boer] are terrible shots, but all the same they came too close for me. One of our fellows [Holloway] got shot right in the head and has been under an operation; a bit of bone was pressing on the brain. Two others were shot through the shoulder, nothing serious to talk about though.[5]

Reginald Holloway. (Droogleever)

Charles Moore was to be killed in action just three months later.

Reg Holloway was admitted to the De Aar hospital where he underwent an operation to repair his skull. The wound to his brain was so serious that it would affect the use of his right leg, and he retained a limp for the rest of his life. While the rest of the Contingent remained in South Africa until late June of 1901, Reg was invalided home to Melbourne arriving on 5 May 1901.

He must have recovered his strength and abilities on the voyage home because a newspaper dated the 7 May reported him visiting his old school.

HOME FROM THE WAR

The boys of Caulfield Grammar School gave a welcome last Friday afternoon, to Private Holloway, an old boy of the school who returned by the *Morayshire* from

South Africa. The local cadet corps turned out in force as a guard of honour to receive him and conducted him to the large hall where he gave some account of his experiences amidst lusty cheers from the boys. In the evening, the Headmaster Buntine entertained several old boys who were at the school with Holloway, to give them an opportunity of meeting and welcoming their old schoolfellow. Private Holloway was seriously wounded in an engagement but is fast recovering strength.[6]

Holloway was later awarded a temporary pension for his injuries and the Boort Branch of the Australian Natives Association gave the family a gold medal:

> To express the esteem in which your son, Mr R. J. Holloway, is held by this branch and we recognise in him a man of true British grit.[7]

This 'true British grit' was being manifested in developments in various recruiting drives that were taking place back in Australia, especially Melbourne in early 1901. Such was the clamour to enlist that the Victorian Government now called for recruits for a 5th Victorian Mounted Rifles contingent. Many men were turned away for a variety of reasons; sometimes too many applied, others had poor health or age limitations, being too old or under the accepted age of twenty-one.

Accordingly, to satisfy the demand for recruits, other 'private' regiments were raised in the British colonies. One such example that enlisted over nine hundred Australians, chiefly from Melbourne, was that of the Scottish Horse, a unit raised by the Scottish Lord, the Marquis of Tullibardine. Advertisements appeared in *The Argus* newspaper of 2 and 6 February 1901 calling for potential recruits to present themselves at Victoria Barracks in St Kilda Road Melbourne for medical, shooting and riding examinations. It was stressed that the candidates must be Scots or of Scottish descent and single, as no married men would be accepted. They should be between the ages of 21 and 28 years, weigh less than 13 stone (82.5kg) and, being first-class riders, they should be accustomed to rough bush life.

Caulfield Grammarians George Raleigh Stewart, George Frederick Roberts and Gerald Massey Ivor Wilkinson were all accepted for the Scottish Horse, along with Malvern Grammarian Henry Gascoigne Davies. It should be noted other Grammarians such as Osmonde Winstanley Birch, Harry James Goodrich Cattanach and Ronald Valentine Sweetwater McPherson also applied to join the Scottish Horse, but were instead accepted into the 5VMR, as were Grammarians Andrew Percival Rowan and Charles Hugh Thomas. For a variety of reasons there was little love lost between the two regiments, highlighted by the fact that the 5VMR 'official government troops' were permitted to parade through the streets of Melbourne to a grand farewell, while the 250 'privately raised' men of the Scottish Horse, all in civilian clothes, were quietly loaded onto their troop transport without any fanfare. The 5VMR contingent comprised forty-six officers and 971 troops and was so large it took three transports to ship all the men, equipment, and horses to South Africa. As the ship left Melbourne on 15 February 1901, to add further insult to injury to all concerned, the Scottish Horse troops had to share their transport with large elements of the 5VMR which must have made for an 'interesting' voyage.

Given the nature of their six-month enlistment, the Scottish Horse soldiers

DISASTER FOR VICTORIA AT WILMANSRUST

The early evening of 12 June 1901 found the 'Left Wing' of the 5th Victorian Mounted Rifles (5VMR) under the command of a British Royal Artillery Officer, Major Morris, camped for the night on a farm named Wilmansrust some 332 km south of the town of Middelburg in the central Transvaal. They were on a mission to find a small Boer force supposedly only 40 km away. Among the 270 or so VMR men were **Ormonde Birch** and **Ronald Mcpherson (F Squadron)** and **Harry Cattanach (G Squadron).** Major Morris chose the position of each sentry post and ordered the men's rifles to be stacked away from their bell tents. At about 7.45pm 150 Boers slipped through the sentry lines and dressed in captured British khaki uniforms fired from the hip as they attacked the camp through the H Squadron lines. It was impossible to tell friend from enemy. Some of the VMR were cooking their evening meal, while others were reading letters or had gone to bed. Pandemonium broke out in the camp and crazed horses broke free and stampeded, knocking over tents. Total confusion reigned and soldiers either froze or surrendered or tried in vain to get their rifles. The fight was over in less than ten minutes, leaving at least 18 Australians killed and 42 wounded. Many men fled into the night, while the rest were easily captured. To add insult to injury, the Boer took all the ammunition, food, clothing and personal possessions from both the living and the dead. Later the blame for the debacle was laid at the feet of Major Morris for his inefficient posting of sentries.

Ormonde Stanley Birch (Droogleever)[9]   Harry James Goodrich Cattanach
                                              (Droogleever)[10]

were not paid until they reached South Africa, when they would also be fitted
out with their uniforms and other equipment. It should also be noted that, in
contrast to the 5VMR and other Australian contingents, the Scottish Horse was a
multi-national force with soldiers coming from all parts of the UK, South Africa,
Canada, New Zealand and Australia.

At this stage of the war and with a move to guerilla warfare by the Boers,
the British authorities increasingly recognised that colonial horsemen were
performing extremely well, had more individuality and were able to find their
way about the country better than their British counterparts. In scouting and
reconnaissance work, the Australians were seen as superior to any troops in
the field.[8]

But before the Scottish Horse took part in any active service, the GHQ at
Pretoria decided to split them into the 1st and 2nd Regiments and send them
to different fields of battle. The 1st Regiment, containing the least number of
Australians, was allotted to Brigadier General Dixon's column of about 1450 men
which, with field artillery support, operated to the north and west of Johannesburg
in the Transvaal. A member of this regiment was Grammarian George Stewart.

The 2nd Regiment, with about two hundred Australians in it, operated in
eastern Transvaal and east of Johannesburg and was allotted to the column of Lt.
Colonel George Benson. It was an accepted fact during this guerilla phase of the
war that, of the sixty or so British columns traversing the South African veldt,

Ronald Valentine Swanwater
McPherson (Droogleever)[11]

Charles Hugh Thomas (Droogleever)[12]

there was none with a finer record than that commanded by Benson. This column was soon seeing action and on 30 April 1901, near the Dullstrom-Roodekrantz district, they encountered a Boer raiding party. In the resulting action, the Scottish Horse's 2nd Regiment suffered their first fatal casualty, a Melbourne soldier was killed, and five other Scottish Horse soldiers were wounded.

As for the 5VMR, when they arrived in South Africa in early March, the regiment was divided into two wings with the Grammarians separated accordingly for nearly all of the rest of their service in South Africa. The 'Right' wing was placed under the command of Major Umphelby and included A Squadron with Andrew Rowan, and C Squadron with Charles Thomas. B and D Squadrons were also part of the 'Right' wing. The 'Left' wing was led by Major McKnight and consisted of F Squadron with both Ormonde Birch and Ronald McPherson and G Squadron with Harry Cattanach. E and H Squadrons

Andrew Percival Rowan (Droogleever)[13]

Elmslie Fayrer Hewitt (https://
alhresearch.tripod.com/2nd_tasmanian_
imperial_bushmen_album/index.
album/170-private-elmslie-fayrer-
hewitt?i=54&s=1)

were also part of the 'Left' wing.

After some early minor engagements, the 'Left' Wing was camped at
Wilmansrust on 12 June 1901 when it
was attacked by a large force of Boers.
Grammarians Birch, McPherson and
Cattanach were fortunate not to have
been among the eighteen dead and
forty-two wounded soldiers. This
was the heaviest number of casualties
of any Australian unit during the
Boer War. From F Squadron (Birch,
McPherson) there were seven killed
and twelve wounded, while G
Squadron (Cattanach) suffered two
dead and eight wounded. To add to the
grief, all the survivors were captured
by the Boers and relieved of almost all
their clothing, equipment, and possessions. Australian military historian Chris
Coulthard-Clark described the 5VMR disaster.

> Many of its members were civilian recruits and in just two months since landing
> in South Africa, the regiment's numbers had been reduced by illness from 1000 to
> little more than seven hundred. After occupying their positions at the Wilmansrust
> farm, sentries had been posted in daylight, making them easily noted and in too
> few numbers to be effective. The VMR's weapons were neatly stacked in piles
> near to where the men slept, rather than right alongside them. The first Boer
> volley, fired from the hip as the attackers ran forward turned the surprised camp
> into a shambles. The fight was over in less than ten minutes. About 50 VMR
> men evaded capture by fleeing into the darkness. The action at Wilmansrust was
> the most serious reverse to befall any overseas colonial force sent to the conflict
> in South Africa, and unfortunately was taken as an indictment of the courage
> and soldierly qualities of the Australian contingents. The lack of vigilance and
> attention to detail concerning sentries rested squarely on the commanding British
> officer, and it was noted that when the Boer attack began, the camp was overtaken
> by a mass panic which defied commanders' efforts to rally the men.[14]

While this costly defeat cast a shadow over the 5VMR, in time the regiment acquitted itself well in future actions and earned one of only six Victoria Crosses awarded to Australians in the conflict.

The former CGS boarder Elmslie Fayrer Hewitt enlisted as a Private (No. 170) in the No. 1 Company of the 2nd Tasmanian Imperial Bushmen (2TIB). This unit consisted of twelve officers and 241 other soldiers, and they sailed with 289 horses on the Chicago from Hobart on 27 March 1901. Numbered among these men was Hewitt's older brother Arthur Reginald Hewitt, who was not a Grammarian. They landed at Port Elisabeth in South Africa on 24 April and served in the Cape Colony for the next 13 months.

The official record stated that:

> Soon after landing the Fourth (Second Imperial Bushmen) Contingent Corps was engaged with Scheepers at Ganna Hoek Cape Colony, where Trooper Warburton was killed. Trooper Brownell distinguished himself in this affair and afterwards received a commission in the Imperial Army. On 19 May, the Tasmanians joined Scobell's Column, which was one of the most successful. On 1 June, they passed to Colonel Gorring, whose force was formed into a Flying Column, without wheeled transport. On 13th February 1902, Colonel Doran took over the Column, and the Contingent served under him until 4th May. On 18th February, they suffered several casualties, and the strain on men and horses was very great; but the Column did excellent work and was frequently complimented by General French and Lord Kitchener. The various leaders commended the Contingent for fearlessness, good horse-mastership, and cheerful endurance of the greatest hardships.
>
> On 18th August 1901, Sergeant-Major Young, of the Cape Police, with Quartermaster Sergeant Lynes, Sergeant Coombes, and eight other Tasmanians, charged a kopje (at Roodeport) where the enemy were strongly entrenched, and captured Commandant Erasmus and others. Young obtained the Victoria Cross. The Contingent were several times successful in capturing influential Boer leaders.
>
> For twelve months they were incessantly employed; long marches often being undertaken by night, followed by actions with the Commandoes of Kruitzinger, Scheepers, Myberg and others. On 22 May 1902, the Contingent embarked on the transport *Manila* at Durban, and arrived in Tasmania on 25th June, having called at Albany, Adelaide, and Melbourne en route. Disbanded on 30th June. (Six members were killed or died during their service.)[15]

Also present in the 2TIB was Lt John Bisdee, who had been awarded the Victoria Cross for gallant actions on previous service in September 1900 with the 1st Tasmanian Contingent. The 2TIB unit's entire service from mid-April 1901 until their departure in mid-May 1902 was undertaken in the Cape Colony and consisted of working as a 'flying column' engaged in pursuing Boer commandoes and capturing several Boer leaders. The 2TIB returned to Tasmania in June 1902 having suffered six deaths from among their ranks, with four from disease and two from action in battle.

Andrew Anderson served with the 4VIB until 22 June 1901 and after his service was completed, he was granted a passage to England. At the conclusion of hostilities in 1902 he was entitled to the QSA with four clasps.

William Boyd enlisted in the 4th Victorian Imperial Bushmen as part of F Company. He was eventually invalided home, arriving in Australia in early October 1900 where he later attended a welcome home to Inglewood and F Company Rangers on 10th December 1900. The local *Serpentine* newspaper, under the heading 'Welcoming Returned Soldiers', reported the following about Boyd on 14 September 1901.

> One of the most successful social gatherings yet held here was that at the Shire Hall last night, when the building was packed with ladies and gentlemen, the occasion being a welcome to Messrs. J. Garden and W. Boyd, returned South African soldiers. The assemblage was thoroughly representative of the district. During the proceedings both guests were made the recipients of handsome gold chains, the presentations, on behalf of the subscribers, being made by Mr W Hardy JP, President of the East Loddon Shire Council, whose remarks were supported by Messrs. W Bissett JP, Ettershank and Cook. The gifts were suitably acknowledged. Dancing was made the order of the evening and was entered into most vigorously.

Edward Duncan was promoted to Lance-Corporal and for bravery in action was Mentioned in Despatches (MiD) for gallantry at Wolmaransstad on 6 March 1901, when he took back spare horses to help to bring out men whose horses had bolted under fire. Promoted to Sergeant, he left for Australia from East London on the *Orient* on 22 June 1901 arriving at Melbourne on 12 July 1901, where he was discharged two weeks later. His war service entitled him to the QSA and five clasps for Cape Colony, Orange Free State, Transvaal, Rhodesia, and South Africa 1901.

His return home was noted in at least two local newspaper accounts.

> Lance Corporal E.A. Duncan of the Australian Imperial Contingent returned to Wangaratta on Wednesday night (17/7/1901) and had a very hearty 'welcome home', some hundreds of townspeople assembling at the railway station to receive

him. The Wangaratta Brass Band played 'Home Sweet Home,' and the young soldier was loudly cheered. He was met by the Mayor, Mr J. Sisley, on behalf of the town and driven to the Theatre Royal where he was entertained, the company including leading residents of the town and district, the local detachment of J Company VMR, rifle club and cadets, Corporal Hennessy and Privates Wallace and H. D. Griffiths. Lance-Corporal Duncan's good health was proposed by Lieutenant C. J. Ahern and enthusiastically honoured and the mayor presented him with a silver tray, liqueur stand and silver entrée dishes. Lance-Corporal Duncan was awarded the medal for distinguished service at an engagement at Wilmanrust, where he galloped out under fire and rescued a member of the Imperial Yeomanry who had been unhorsed.[16]

The same newspaper article then outlined their return to their homes.

On Thursday (17/7/1901) Lance Corporal E. Duncan and Troopers A.D. and A.H. Wallace of the Imperial Bushmen's Contingent returned to their homes at Wandiligong. At Bright about five hundred people were present and the soldiers were formally welcomed by Councillor Maddison on behalf of the residents. A large four in hand coach was then boarded and with a brass band playing, a procession about a half mile long proceeded to Wandiligong where a hearty welcome was given.[17]

An article published the following year to mark the near end of the Boer War described the scene.

An outburst of enthusiasm, only once before equalled in this locality (Wandiligong), and that was on the return from the war of Lance-Corporal Duncan and Troopers A.H. and A.D. Wallace. Messrs A.H. Wallace and E.T. Duncan were also on the platform and when during the evening Trooper Forrester of Bright, arrived and was welcomed by the chairman, his appearance was the signal for another outburst of cheering and acclamation.[18]

While some soldiers returned home, others were preparing for a second tour of duty.

Endnotes

1 King. J. *Great Battles in Australian History.* Allen and Unwin. Sydney. 2011. p26.

2 Drooglever. R. *That Ragged Mob – Being the Service Record of the 3rd and 4th Victorian Bushmen Contingents in the Boer War.* Trojan Press. Melbourne. 2009. p164

3 https://www.awm.gov.au/collection/ U52011

4 Murray, P.L., *Official Records of the Australian Military Contingents to the War in South Africa.* 1911.

5 Mann. J. and Allen D. *Fallen. The Ultimate Heroes. Footballers who never returned from war.* Crown Content. Melbourne. 2002. p5.

6 nla.gov.au/nla.news-article10555156

7 Drooglever. R. Personal email to Dr Daryl Moran in October 2019

8 Australian War Memorial: *South African War 1899–1902.* Campaign Series No. 2. Progress Press Canberra. p5.

9 Drooglever. R. *A Matter of Honour. Being the history of the 5th Contingent of the Victorian Mounted Rifles in the Boer War, 1901–1902.* Trojan Press. Melbourne. 2017. p492.

10 Drooglever. R. *A Matter of Honour.* ibid. p494

11 Drooglever. R. *A Matter of Honour.* ibid. p507

12 Drooglever. R. *A Matter of Honour.* ibid. p515

13 Drooglever. R. *A Matter of Honour.* ibid. p512

14 Coulthard-Clark C. *The Encyclopaedia of Australia's Battles.* Allen and Unwin. Sydney. 2001. pp91–2.

15 Murray, P.L., ibid. pp. 561–562.

16 *Argus.* 22 June 1901. p5

17 *Argus.* 22 June 1901. p5

18 *Ovens and Murray Advertiser* (Beechworth) 10th May 1902. p4

CHAPTER 7

PHASE 3 (PART 2) –
FINAL BRITISH VICTORY
SEPTEMBER 1900 – MAY 1902

In early 1901, other Grammarians were joining the fray in South Africa, some for the second time including Stanley Reid, who had been offered a lieutenant's commission. Reid, with his brother Dr Francis Reid and another Caulfield Grammarian, Lt Bernard Bardwell joined the other officers and 240 men of the 6th West Australian Mounted Infantry (6WAMI) which left Fremantle on 10th April 1901, arriving in Durban on 29th April. Attached to Major-General Kitchener's column of about 1400 men, 6WAMI had joined with 5 WAMI and marched south-east out of Middelburg on 13th May. It was at the settlement

Lt **Bernard Bardwell** is seated on the ground at left in this photo of the officers of the 6th WAMI. It is undated but might have been taken to mark the death of Queen Victoria on 22nd January 1901, as they are all wearing black mourning arm bands on their left arms. Standing at the left is another Caulfield Grammarian, **Lt Stanley Spencer Reid** who was killed in action on 23rd June 1901. His older brother, the Unit's Medical Officer, Surgeon Captain Francis Reid who treated Stanley for his mortal wound, is seated at the far left. Lieutenant Frederick Bell (standing far right) became West Australia's first Victoria Cross recipient for bravery in action on September 1901.[1]

AUSTRALIAN VICTORIA CROSS WINNERS

The Victoria Cross was instituted by Queen Victoria in 1856 to honour, '*the most conspicuous bravery,*' the award taking precedence over all other decorations and medals. During the Boer War 78 members of the British armed forces, including six Australians received the medal. **Lt Neville Howse** served with the NSW Army Medical Corps and rescued and treated a wounded soldier under heavy fire. **Trooper John Bisdee,** a member of the Tasmanian Imperial Bushmen rescued a wounded officer, also under heavy fire. **Lt Guy Wylly** was also a member of the Tasmanian Imperial Bushmen who was leading a scouting party when he was shot and wounded. He saved a wounded man and then put up an armed defence to save his party from death or capture. **Lt Frederick Bell** was a member of the West Australian Mounted Infantry,

who when retiring from a position through heavy fire, returned on his horse to save a wounded man. The horse could not carry both men, so Bell remained behind and provided cover to allow the soldier to escape. **Sgt James Rogers** had served in the Victorian Mounted Infantry, but at the end of his service with them, joined the South African Constabulary. Under heavy Boer fire, Rogers, on horseback, rescued all members of his small party of ambushed six soldiers from a much larger attacking Boer party. **Lt Leslie Maygar** was a member of the 5th Victorian Mounted Rifles, who galloped out under fire to order an outpost to retire to safety. A man's horse was shot from under him, so Maygar dismounted and gave his horse to his comrade, enabling him to escape. Under ongoing heavy fire, Maygar eventually escaped to safety on foot.

of Vlakfontein on 16th May 1901 in an engagement with a party of Boer, that Stanley Reid was first wounded. The event was recorded in the Official History of the 6WAMI as follows:

> In extended order they advanced, Lt Reid taking his division of approximately twenty troopers to the extreme left. Here he had to encounter a very heavy fire from a mealie (grain) patch on the right of the farmhouse and was shot through the abdomen, but gallantly continued to lead his men.[2]

Out on the right flank, a brother officer, Lt Frederick Bell's gallant actions in saving a wounded man under fire, earned him the Victoria Cross. In a letter received in Perth on 8th July, Reid wrote of his wounding.

Here I am, living in a field hospital, with a bullet through part of my stomach. It was a week ago that it happened, and I have had an extremely fortunate escape. The bullet entered my left side, just above and a little to the rear of my hip bone and came out in the pit of my stomach, by some miraculous means missing my intestines. The good luck was evidently due to my position over the horse's neck. Altogether it was a disastrous day for the Westralia. Eight deaths have occurred already, five in ours and three including Lt Forrest, in the Fifth. Vlakfontein was the scene and without a doubt the honours rested with the Boers who fought splendidly with head and heart. I cannot help admiring them and only long to cross swords with them again to see if we cannot restore the balance, which at present, is in their favour. How well the Boers shot that day, and they were as cool and steady as rock. Besides the killed, five or six were severely wounded and as many taken prisoner, stripped of all they had and turned loose. I feel awfully sorry for the relations of the killed in Western Australia. Alas! It means something more to be wiped off the slate. I feel as jolly as a sandboy; just a little impatient at the confinement for the doctors will not let me walk, although I can do so with comparative ease. After all is said and done there is something most enthralling in a soldier's life; probably the excitement of uncertainty.[3]

Stanley Reid had recovered from his wound well enough to be in action again a month later on June 23rd, when 5 and 6 WAMI carried out a reconnaissance in force near Spitzkop, at a place called Tweefontein, some 15 miles north-west of Ermelo. Stanley Reid's division was in line at some kraals (cattle enclosures) when two Boers appeared on the skyline and leaving four or five men, Reid and the rest of his small troop gave chase. They galloped over the crest and out of sight and leaving the horse-holders, eleven of the party advanced on foot and walked into a well-prepared ambush.

A letter written by his brother Surgeon-Captain Reid published in August the Western Mail outlines the tragedy that then unfolded.

At a range of only about fifty yards, the fire was sudden and terrific and must have dropped the two who were a bit to Lt Reid's right and who were killed outright. Reid's sergeant, William Mills, was wounded three times before two men managed to run back to the main force which then directed artillery fire onto the Boer position. After they had finished, my orderly and I rode across to the scene carrying the Red Cross flag. There we found two dead and three wounded – Stanley through the stomach, his sergeant in three places, through the neck, through the leg and through his lungs, and a private through the chest.

After dressing their wounds, in about one ½ hours, the ambulance wagon came and took them back to camp.[4]

Fellow 6th WAMI officer and Caulfield Grammarian, Lt Bernard Bardwell also wrote about the incident in a letter published at home in August 1901.

Stanley Reid was shot in the stomach again, and one of the Boers who was present when he was shot the first time, remarked to his brother (the doctor), 'Well we have got your brother again.' Sergeant Mills after being wounded three times said as he was being moved off the field, 'They have scored two centres and one outer, but they have not got a bull's eye yet.' He felt OK until the ambulance came, and he was put into a cart with Mr Reid, and they had a very bumpy journey back to the camp by which time he felt bad.[5]

Reid's brother, Surgeon-Captain Francis Reid tended to his wound and attended him constantly and described the circumstances in a letter to their parents.

Stanley died [on the morning of 29 June] at 5 o'clock. Since he was wounded, I have been near him all the time and was with him when he passed away. I had all the surgeons in the camp, but they all agreed it was hopeless. He was in considerable pain but stood it as I have seldom seen a man stand it. The men of the contingent fairly worshipped him and are very cut up over his death. Poor Stanley's grave is the best I have seen in South Africa. The men asked leave from the captain to look after the grave. Leave was granted, and they worked away at it and made it up splendidly.[6, 7]

Lt Bernard Bardwell commented:

Poor Stanley Reid died this morning. We buried him alongside some of the Victorians who were killed. His brother, the doctor was almost mad with grief. It will take a strong hand to pull him together again, as he is utterly broken down, poor fellow.[8]

After hostilities ceased in May 1902, Reid's body, along with others was reburied at Middelburg Cemetery.

Bardwell and the remainder of the WAMI returned to Middleburg and then moved further into the Eastern Transvaal to act for the rest of 1901 under the command of General Hamilton. Light casualties were sustained as the Contingent undertook many arduous marches in pursuit of the Boer, but to little avail. Eventually both Contingents embarked for Australia on 7th April 1902 arriving at Fremantle on the 29th and being disbanded on 17th May 1902. The contingent had suffered ten men killed or died of wounds, including Stanley Reid and four deaths from disease.

Due to contracting enteric fever late in the campaign, Bardwell [who later recovered] had been hospitalised and was then repatriated later than the main Contingent, as he received a letter at Perth on 15 May 1902 written by a former member of his unit.

Dear Sir, I write on behalf of the non-commissioned officers and men of No.1 Division 'B' Company of the Sixth West Australian Mounted Infantry. I wish to express to you their appreciation for the very capable and pleasant manner in which you carried out the duties and responsibilities as their divisional officer, whilst on active service in South Africa, and I beg to hand you herewith a framed photo of the members of the division, who returned to Western Australia. Our only regret is that you were not able to be present to complete the group. Trusting that you are now thoroughly recovered [from an attack of enteric fever] and wishing you every success. I remain, yours very faithfully, H.E. Blumenthal. (late Private Sixth WAMI)[9]

One of the Scottish Horse recruits who been in South Africa since February 1901 was Gerald Massey Ivor Wilkinson, known to all as Ivor or 'Jinks'. Not long after leaving CGS, Ivor gained a pharmacy apprenticeship with his older brother Paul in their late father's pharmacy in Maryborough, but it was stated that Ivor found this to be a distasteful occupation, as he found that indoor life did not suit him at all.[10] When the Boer War broke out in 1899, he volunteered with the 5VMR for active service, but was rejected due to his young age. In 1992, his great nephew Peter Ross later recalled being told that 'Jinks' had joined with two other close friends, and that they had ridden their bicycles 50 km from Maryborough to Castlemaine in an attempt to enlist.[11] It is salutary to remember the mood of Australians at the time as recalled by Ivor's old school friend Reginald Greenwood.

I can recall the wave of patriotic fervour which swept over Victoria and other states. Thousands of men from the cities and outback made their way to the recruiting depots and Ballarat in particular mustered strong contingents. What exciting days they were![12]

In November 1900 Lord Kitchener sanctioned the raising of a regiment to be known as the Scottish Horse. Lord Tullibardine soon started recruiting from Scotsmen, or men of Scottish descent, in South Africa, chiefly in Natal; and on 4th February 1901 he took the field with three squadrons. To these other

The Marquis of Tullibardine (left) and another officer ride past soldiers waiting beside railway carriages. (AWM P01115.013)

squadrons were soon added others and the Volunteer Service Companies of Scottish regiments furnished no less than two hundred men. To these men their leader, Tullibardine, gave the highest possible praise. 'One hundred of them were the best body of men in every way that I saw in South Africa. This particular squadron had a reputation which extended far beyond the column with which it was trekking.' Recruiting was not confined to South Africa and Great Britain and the other Colonies were appealed to, and the Caledonian Societies in London and overseas did grand work. The Highland Society of London sent out 386 officers and men, who sailed in February and March 1901; and the Marquis's father, the Duke of Atholl, personally raised 831 men before the war was over. The Society in Melbourne took up the matter with enthusiasm, and 'about 300 men joined on 8th March and were a splendid draft, very fine riders, and all Victorians'. Later, more men joined from Australia, recruiting having been attended with success. The first regiment was soon six squadrons strong, and a second of five squadrons also took the field.[13]

The historian of the Scottish Horse, John Price noted that the military authorities in Australia at the time were decidedly cool towards the raising of troops for the Scottish Horse, as it was seen to be detracting from recruiting for the official State contingents. In fact, the Victorian Government on the 4th of January 1901 edition of *The Argus* newspaper firmly stipulated:

> … that should any volunteer for the Scottish Horse get killed or maimed through service his family could claim no local compensation.[14]

Nevertheless, all the seven hundred or so applicants were eventually weeded out through a medical examination and those under the age of twenty-one were deleted from the process. Ivor persevered with his desire to enlist by travelling to Melbourne and, despite only being seventeen, passed himself off as being much older and was enrolled (no. 31757) in the 32nd Regiment of Tullibardine's Scottish Horse. Candidates also had to undergo riding tests on the Domain and musketry tests at Elwood rifle range. In addition, candidates had to prove their Scottish ancestry, something that Ivor, with his English parentage must have also shown some 'creativity' in doing. Eventually some 250 men were approved, Wilkinson included, and they signed on under the following conditions as outlined by Price.

> The men were then signed on as members of the Scottish Horse, on the understanding that:
>
> 1. They agreed to serve six months from the date they were sworn in at Cape Town.
> 2. Uniforms and accoutrements would be issued at Cape Town.
> 3. Pay was to start from the date of swearing in, at the rate of five shillings a day.
> 4. They would be discharged and paid off at Johannesburg at the end of six months.[15]

As events transpired, the men were paid from their embarkation from Melbourne, but still had no guarantee of a paid passage home at the end of their service. Ivor Wilkinson sailed to South Africa on the *Orient* on 15 February 1901 and as a member of the fifth contingent was posted to F Squadron. In a buoyant mood and once in South Africa he wrote a happy letter home, of his early experience:

> I celebrated my 18th birthday on the 12th of May by having a jolly good tuck-in of tinned sausages and stewed fruit and by way of dessert, I received a pleasant surprise in the form of lance-corporal's stripe, with promises of a corporal's stripe.

The Scottish Horse were posted to the Transvaal area to the north of Johannesburg where British General Sir Bindon Blood was in command of the East Transvaal region and one of his senior cavalry commanders was Lt. Colonel George Benson. This large, mounted formation, which now included the 2nd Scottish Horse, had the job of methodically clearing their assigned area of Boers, which they did with increasing success, capturing many prisoners and supplies in the process. But sadly for 'Jinks' the outcome of gaining corporal's stripes was never to be, as his fate was outlined in an official history of the Regiment as it briefly told of its second engagement with the Boer forces.

> Blood's July operations began as early as the third when Benson left Machadodorp and headed towards Dullstrom to attack an enemy laager (camp) at Vlakfontein, also known as Elandskloof, which was midway between the two towns. The enemy, having been alerted, feigned flight before turning on the Scottish Horse advance guard. In the close combat which followed Lance Corporal 'Jinks' Wilkinson, Troopers Clay and Whiting, all Victorians, were killed and nine men wounded, including five Australasians. In return the Boers suffered twelve casualties. For bravery in this action Lt. William English was awarded the Victoria Cross.[16]

The battle was further described in a soldier's letter home published in a newspaper and headed, 'Victorians Killed.'

> Trooper G. Crawford of Prahran, who is a member of the Second Scottish Horse in South Africa, describes in a letter to his father, a 'hot' fight with a Boer commando, under (its leader) Pretorius. 'It lasted a couple of hours when reinforcements arrived and the Boer were forced to retreat and Lance Corporal Wilkinson of Maryborough, was killed when struck just below the ribs by an explosive bullet.[17]

A closer examination, however, reveals a slightly different story and an article printed in the 1994 South African Military History Society publication *Military History Journal* gives more detail to the circumstances surrounding Ivor Wilkinson's death as a member of F Squadron. An extract from the diary of Sgt R B Hodgson of E Squadron 2nd Scottish Horse outlines the engagement.

> We had been in tents ever since we had been in Machadodorp, but we were ordered to pull them down that night and, worse still, reveille was at 01:00 and we were on the march half an hour later and, after going about twenty miles [32 km] halted for dinner. We had left the convoy with the Argyll [and Sutherland Highlanders] some distance in the rear and, with the aid of our glasses, could distinguish the place of our first fight, about three miles [4.8 km] away. It was

3 o'clock in the afternoon before the order was given for E Squadron to retire to camp. The other squadrons were already halfway there, except for F Squadron, which had been divided into two detachments of thirty-eight men each and sent to search two valleys some distance apart. There appeared to be no danger in separating the squadron at the time for there were no Boers in sight. We had gone only two miles [3.2 km] when firing was heard, and a messenger galloped up to say that part of F Squadron was surrounded by Boers. Captain Kelly [his squadron commander] shouted 'Files about, gallop!' so off we went back to their assistance, over rocks, ditches, all sorts of obstacles. Every so often a horse would 'turn turtle' and send his rider a header. We galloped right round the flank of the burghers and dismounting, started to advance. The chaps of F Squadron told us later that when they saw us, they thought it was all up with them for they had mistaken us for another party of Boers. The enemy was so intent on the capture of the squadron, that we held a good advantage before they saw us. Then they tried to keep us off, but the captain, firing and advancing in short, sharp rushes, showed that he meant business. We blazed away at each other for some time and then our pom-pom (gun) came up and started to pump shells into them. That settled it, for the Boers broke and left us in sole possession together with six of their dead. Our losses were heavy, mostly F Squadron men, Clay, Whiting, and Wilkinson, all splendid young fellows. Wilkinson, well-liked, was killed attempting to silence a Boer who was firing on the horse holders, having crept to within ten yards [9 m] of his prey, when his rifle jammed, and he was shot through the brain. Eleven others were wounded but recovered afterwards. It would have gone very hard for F had it not been for our timely arrival, for they had expended nearly all their ammunition, and, in some instances, the Boers were within fifteen to twenty yards [14–18 m] of them. After burying the dead, we returned to camp with the ambulance wagons taking our wounded into Machadodorp.[18]

Ivor's death was reported in newspapers in a variety of ways, with one report under the heading 'The Disaster to Tullibardine's Horse' detailing some circumstances in Maryborough.

At the state school today, Mr Carter, the head teacher, read to the assembled scholars *The Argus* cable message, announcing the death of Ivor Wilkinson of Tullibardine's Scottish Horse, in South Africa, a former pupil of the school. Mr Carter made touching reference to the sad event and expressed sympathy with Mrs Wilkinson, the deceased soldier's mother, and his family in the loss they had sustained. He spoke of the bright and genial disposition of Wilkinson, who was always a favourite amongst his fellows and who was the first of Maryborough's boys to give his life for the honour of his country and the flag in South Africa. The Union Jack was lowered to half-mast out of respect to the memory of the deceased and with bared heads, the boys saluted the flag before marching into their respective places. The flag at the town-hall was also flying at half-mast. Mr Paul Wilkinson, chemist of this town, only this week received a letter from his brother, in which he stated that he had been promoted to lance-corporal and intended to engage for service for a further term. The deceased was only 18 years of age. The Rev. Canon Harris has arranged to hold an in-memoriam service in the town-hall on Sunday evening, which will be attended by the borough councillors, the Mounted Rifles, the rifle club, and the cadets.[19]

His death notice pointed out that he had been:

Killed at Elandshoek on July 3rd, 1901, while fighting, true to the spirit of the words which caused him to be named after the poet (Gerald Massey) who wrote the following verse.

> Old England still hath heroes
> To wear her sword and shield:
> We know them not while near us.
> We know them in the field:
> But there's no land like England,
> Wherever that land may be.
> Of all the world 'tis King-land,
> Crowned by its Bride the Sea.
> And they rest 'I the balmiest bed,
> Who battle for it, and bleed for it:
> And they shall be head of the glorious dead,
> Who die in the hour of need for it.

The official program for his memorial service also poignantly noted that, 'his parents had such a heartfelt admiration for the verse from a poem of Gerald Massey's, that they named their youngest son after that author, little thinking that their boy's fate would so nobly fulfill the spirit of the words.' Ivor Wilkinson's memorial service was conducted on Monday 15 July 1901 in the Maryborough Town Hall and was attended by an enormous number of people with many being unable to obtain admission. The Rev. Canon Harris of Christ Church conducted the service and preached the sermon. The Maryborough Borough councillors, mounted rifles, rifle club, cadets and other public bodies attended and evidenced the respect in which the deceased soldier was held. [Much] sympathy was extended to his mother and relatives.[20] The final words about the death of Ivor Wilkinson were left to Lord Tullibardine himself, who wrote to Ivor's mother from South Africa just three days after the fatal action.

I feel very sorry to have to a send to you the enclosed official notification [of his death] and I know it will be a great blow to you. Your son did so well out here, too. We must remember, however, that he died well for his country, and you must be very proud of him – not sorry. I will send you more particulars when I get the despatches. I saw him on the morning of the 3rd, just before he went out as I was going down the line myself. I fear my sympathy cannot help much, but you

BATTLE OF BAKENLAAGTE

Lord Tullibardine's multi-national mounted Scottish Horse had been divided into the 1st and 2nd Regiments and Grammarians **Henry Davies** and **George Roberts** of the 2nd Regiment, had been allocated to the Mounted Column of Lt. Col. George Benson. On 30 October 1901 this column was located on a farm at Bakenlaagte, some 35 km north-west of Bethal in eastern Transvaal. The formidable Boer Commander Louis Botha, with a mobile force of 2000 men saw a chance to inflict serious damage on the British column, as the 950 mounted men and 650 infantry soldiers began their slow march towards a re-supply base. The Scottish Horse, including Davies and Roberts, dismounted and with the infantry withdrew to a better defensive position on slightly higher ground. They were ordered to hold the ridge and to save the guns, so the British troops formed a straggly line on both sides of the two guns to defend them. The Boers moved relentlessly forward with deadly gun fire and in time killed all the gun crews where they had stood. The Boers then surrounded the position at close range and continued to pour in a murderous fire on all the troops. By this time Benson and his senior officers were either dead, mortally wounded or seriously wounded. The sheer weight of numbers and British casualties meant that the mass of Boers finally captured the two guns and dragged them away. They then stripped the British bodies of their clothes, boots, arms, and valuables. Total British casualties amounted to 238 men killed or wounded and 120 captured. Scottish Horse casualties were 33 killed with six of them being Victorians. Davies and Roberts were both unhurt.

know I mean it, all the more as he came out to serve under me. Yours faithfully, Tullibardine.[21]

In June 1901 Frank Weir and the NSWCBC had completed their tour of duty and had returned home as had Andrew Anderson, William Boyd, Edward Duncan and George Williams when the 4VIB returned to Melbourne in the same month.

While the Battle at Elandshoek on 3 July 1901 where Ivor Wilkinson had been killed was viewed as a major skirmish, a much grimmer test awaited the 2nd Regiment of the Scottish Horse. Australians involved included Malvern Grammarian Henry Davies and Caulfield Grammarian George Roberts. On Wednesday 30 October 1901, a large Boer force attacked the column of Lt. Col. George Benson at Bakenlaagte, which included a rearguard of the Scottish Horse who valiantly attempted to save the artillery pieces of their support unit. Two hundred and forty British troops were killed with thirty of them being Scottish Horse. Those numbers included ten Australians killed. Davies and Roberts fortunately, were not among the dead and wounded.

The year of 1902 brought more Grammarians to the conflict in the first Australian units to be deployed after the Australian colonies had federated on 1 January 1901. One soldier was serving for the second time, while another Grammarian was wounded in action early in the year. George Raleigh Stewart enlisted on 21st December 1901 as a 26-year-old Trooper (36892) in the 1st Scottish Horse and was severely wounded at Gruisfontein on 5th February 1902 as recounted below.

In the early months of 1902, the regiment was constantly on the trek and fighting. At Gruisfontein, on 5 February 1902, the whole of Sarel Albert's commando was captured. As to this action, Lord Kitchener, in his despatch of 8 February, said: 'During Major Leader's advance he came upon and captured a Boer picquet, from which he ascertained that General Delarey had already moved his camp, but that Commandant Sarel Albert's laager was for that night at Gruisfontein, which he reached just before daybreak. Our men charged the enemy's laager with great dash, the Scottish Horse taking the main share of the attack, and as most of the Boer horses had been stampeded by the fire of Major Leader's pom-pom, the gallantry of the attacking force was rewarded by an unusually large measure of success; 7 Boers were killed, 132 prisoners taken, 11 of whom were wounded, together with 130 rifles, 2800 rounds of ammunition, and a large number of horses, mules, cattle, and wagons were taken. Our casualties were two officers (Captain Ian R McKenzie and Lieutenant W Tanner), and six men wounded, including Trooper George Stewart, all belonging to the Scottish Horse.' In his telegram of 5 February Lord Kitchener said: 'Leader reports that the Scottish Horse behaved with great gallantry'.[22]

George returned to Australia and was discharged on 10th October 1902 and was entitled to the QSA and three clasps for service in the Transvaal and South Africa (1901) and South Africa (1902).

Members of the 1st Battalion, Australian Commonwealth Horse having a rest break and a meal. (AWM P04871.002)

Later in February 1902 after treatment and convalescence, Frank Weir returned to Australia and promptly enlisted in the 1st Battalion of the newly formed Australian Commonwealth Horse. This was a mounted infantry unit of the newly constituted Australian Army formed for service during the Second Boer War in South Africa in 1902. It was the first expeditionary military unit established by the newly formed Commonwealth of Australia following Federation in 1901. Battalions of a representative character were formed, with squadrons from different States with the first made up of three units from New South Wales, one from Queensland and one from Tasmania. The men selected were required to be good shots and good horsemen; men of previous experience having preference, if medically fit. Only single men were taken, and unmarried officers had preference.

The 1st Australian Commonwealth Horse (ACH) (New South Wales units) with twenty-one officers, including Weir and 354 soldiers embarked at Sydney on the transport *Custodian* on 18 February 1902, and disembarked at Durban on 19 March 1902. They eventually joined the Column under the command of Colonel De Lisle, which formed part of Thorneycroft's brigade. The Column was employed clearing the district north of Klerksdorp and took part in a drive which commenced on 7th May, moving westward to the Kimberley–Mafeking railway blockhouse line. In this, 251 prisoners, including General De la Rey's brother, were captured, also three hundred horses, 144 riders and bandoliers, 6,000 rounds ammunition, and a large quantity of stock. Colonel De Lisle, who was leaving for England, handed over his command to Colonel Williams. The Column then returned to Klerksdorp, reaching there on 21 May, where the regiment remained until the declaration of peace, when they were ordered to Elandsfontein to prepare for return to Australia. Having seen only two months of active service, the 1st ACH were luckier than many other units which were still sailing for South Africa when peace was declared. Twenty-one officers and 330 others returned by the transport *Drayton Grange* which left Durban on 11th July, and arrived at Sydney on 11th August 1902, having called at Albany and Melbourne on the way. Frank Weir's two periods of active service were acknowledged by the awards of the QSA and four clasps and the KSA and two clasps.

James Campbell had already served in South Africa from April 1900 to May 1901 with the 6th Imperial Bushmen from NSW. Named on his enlistment papers as 'a gentleman aged 25 from Brighton,' he now enlisted as a Lance Corporal (2275) in the 4th Australian Commonwealth Horse. This unit sailed for Durban, South Africa on 26th March 1902, unknowingly while peace negotiations were underway. When they arrived on 22nd April they were posted to the Transvaal and saw no action, eventually returning their equipment and horses when peace was declared on 1st June. The unit returned to Melbourne later that month.

A champion athlete at Malvern Grammar School was the last Grammarian to join the fray. Charles Reginald Handfield was born in South Yarra on 26th August 1878 and was the youngest son of six children of Mary and Lt Frederick Oliver Handfield who had served in the Victorian Volunteer Navy from 1861–1870. The exact years Charles was at Malvern Grammar School are uncertain, but he was certainly a student during the years of 1895–96 as he was named captain of both the football and cricket teams as well as being named the School [Athletics] Champion. In 1896, when he was aged eighteen, there were seventy-three boys on the roll at MGS and it was a notable year in the school's history as in addition to success in study, there was also good success in sport. The football team won all its matches, as did the First XI which notably had a draw with the Second XI of Scotch College. Charles' prowess was further illustrated in a letter from another Old Boy to the school magazine *The Malvernian* in July 1907.

> The event which I would like to notice was 'Throwing the Cricket Ball' at the 1896 sports, and I think you will find the following particulars correct. The event took place on the vacant land opposite the Church and was won by C.R. Handfield with a throw of one hundred yards (91.4 metres), Mr McLean [Headmaster] pacing the distance immediately after the finish. The ground was carefully marked and the following morning the distance was confirmed by a tape measure. The record is one which will not be beaten easily and should certainly be kept among the other records of Malvern Grammar School boys.

When hostilities broke out in South Africa in late1899, several volunteer regiments were formed and chiefly recruited from among local men, especially in what could be termed the 'ex-patriate' community or 'Uitlanders'. In the Afrikaans language, this term meant a foreigner, which applied to any British or other non-Afrikaner immigrant in the Transvaal; regions of South Africa in the 1880s and 1890s. In many cases these volunteers were men who had been resident in the Transvaal and other areas but had kept their allegiance to Britain. One such unit was the Imperial Light Horse (ILH) which was formed in late 1899 and came to be regarded as one of the best units in the field. It saw service at the siege of Ladysmith, Colenso, Spion Kop and Relief of Mafeking. They also undertook campaigns in the South African Republic and the Orange Free State until the end of the war. An account of the ILH reads:

> Further successful drives took place in the south-west Transvaal in which the ILH were engaged. Thus, from Elandslaagle to the very last stage of the war, did this splendid Volunteer Regiment keep steadily at work. Throughout the whole war they had done nobly; no regular troops could have reached a higher standard and if they were largely men who had stakes in the Transvaal, they did all they could to assist the mother country in the struggle for the maintenance of British sovereignty in South Africa.[23]

Thousands of men volunteered to serve in the ILH, and they were subjected to a rigorous selection process. The men chosen were the best in horsemanship, shooting, character, and physique. The training that these men received was equally rigorous and those lacking the necessary keenness and efficiency were discharged and replaced by new more suitable recruits. About 45% of the men were South African born, 45% were British by birth, while the rest were from other Colonies (Australia, New Zealand, and Canada) and countries elsewhere in the world.

After leaving school Charles Handfield and his brother George moved to Johannesburg in South Africa in 1901 where on 25 February 1902 Charles enlisted in the 1st Imperial Light Horse (ILH). Handfield was promoted to be a Corporal and having completed four months service with the ILH, in the Orange Free State, received the South Africa (service) Medal in 1902 and was also entitled to wear clasps for active service in the Orange Free State and the Transvaal.

Gradually the Australian soldiers returned home and on 25 and 26 April 1902, 5VMR including Grammarians Ormonde Birch, Harry Cattanach, Ronald McPherson, Andrew Rowan, and Charles Thomas returned to Melbourne. Given the tragedy at Wilmansrust, it is not surprising that the unit's total losses amounted to thirty-six men killed or died of wounds and thirteen from disease. But the end was in sight for both weary sides.

> By the Third year of the War the position of both sides on the question of peace was shifting. Boer and Briton were being worn down by a guerilla campaign fought over enormous distances. Boer strength and resources were dwindling and their families still on the veldt were close to starving. Kitchener's British [long serving] troops were also approaching exhaustion.[24]

Accordingly, and after various negotiations between both sides, the Treaty of Vereeniging was signed on 31 May 1902 and the Boer commando finally laid down their arms and surrendered.

What had been the cost? At the end of the war in 1902, it was estimated that over the three years of the conflict, the Boers lost some six thousand military casualties, but a further 24,000 Boer prisoners were sent overseas into exile. Britain had sent some 450,000 troops to fight about 50,000 Boers and that it had cost over $350 billion (AUD) in today's money terms. Britain's total casualties were 22,000 men with over 14,000 who died from disease, principally enteric fever.[25] It is believed that Australia lost over one thousand soldiers with some 280 falling

Charles Reginald Handfield.

victim to disease. Over eight thousand British soldiers were battle casualties with Australia losing close to three hundred dead in action.

Twenty-four Caulfield Grammarians and four Malvern Grammarians served in

---

'TO END A WAR: THE TREATY OF VEREENIGING, 31 MAY 1902'

Excerpts from an article by Professor Peter Stanley

'By early 1902, the Boer republics were occupied and their people interned. Exhausted Boer commandos were still in the field as 'bitter-enders,' but they were unable to achieve victory. Boer leaders debated how they could end their suffering and retrieve a peace they could live with. Accepting defeat, they still hoped to retain a measure of political and cultural autonomy and opened negotiations with their British counterparts. While some British figures wanted the unconditional surrender and the elimination of the Boers' Afrikaaner language, the Boers' negotiators largely gained what they sought. Britain was also tired of the largest conflict it had seen since the defeat of Napoleon, and they wished to see the war end. [The meetings took place in the small Transvaal town of Vereeniging, but the treaty was signed in Pretoria.] The treaty's main points were that all Boer combatants would surrender, be temporarily disarmed and swear allegiance to the British crown. Consequently, all these Boers received an amnesty and returned home to their (often devastated) farms, though they were eligible for small reconstruction grants. The war ended the Boer's independence and the two republics [of Transvaal and the Orange Free State] became part of the British empire. The Boer continued to use Afrikaans in schools, churches and courts and while at first these were administered by British officials, the republics were expected to regain self-government in a British South Africa, which was achieved within five years.'

https://www.anzacmemorial.nsw.gov.au/ our-stories/our-stories/end-war-treaty-vereeniging-31-may-1902

---

the Boer War. Three Caulfield Grammarians were killed in action. Four Caulfield Grammarians died of disease. Seven Caulfield Grammarians paid the supreme sacrifice. How were these deaths marked by Australia, their local communities and for our special interest, their former secondary school, Caulfield Grammar?

Endnotes

1 http:// www.bwm.org.au/contingents.php

2 Campbell. J A. (Lt. Col.) (ed) *History of Western Australian Contingents serving in South Africa during the Boer War (1899–1902)* (HWAC) Government Printer, Perth, 1910. p43

3 Letter published in the *Western Mail.* 13 July 1901. p44.

4 Surgeon-Captain Francis Reid. Letter in the *Western Mail* 17 August 1901. p47

5 Lt Bernard Bardwell. Letter in the *Western Mail.* 31 August 1901. p44

6 The Late Lieut. Stanley Reid: Letter from His Brother, *West Australian*, Tuesday 13 August 1901, p.6

7 The Late Lieut. Stanley Reid: Letter from His Brother, *West Australian*, Tuesday 13 August 1901, p6

8 Lt Bernard Bardwell. ibid.

9 nla.gov.au/nla.news-article33212201

10 MacKay S. *Advertiser* Friday 5 July 1991. p11

11 Ross P.M. Letter to D.J. Moran. 13 December 1992

12 Greenwood. R. ibid.

13 https://www.angloboerwar.com/?option=com_content&view=article&id=469

14 Price. J.E. *Southern Cross Scots.* BR Printing. Kensington. Vic. 1992. p31.

15 Price. J.E. ibid. p32.

16 Price. J.E. ibid. p54.

17 *Argus.* (Melbourne) 10 September 1901. p6

18 http://samilitaryhistory.org/vol096jh.html

19 Undated newspaper clipping held in CGS archives.

20 *Argus.* Tuesday 16 July 1901. p6

21 *Argus.* Thursday 8 August 1901.p5

22 https://www.bwm.org.au/soldiers/Hobart_Cato.php

23 https://angloboerwar.com/unit-information/south-african-units/345-imperial-light-horse

24 Smurthwaite. D. *The Boer War 1899–1902.* Hamlyn. London. 1999. p169.

25 www.familyhistory.co.uk/the-boer-war pp5–6.

PART FOUR

THE BOER WAR
AFTERMATH,
WAR AND MEMORY

CHAPTER 8

GRAMMARIANS DEATHS REMEMBERED

B ritain sent 450,000 men to South Africa in 1899–1902, losing 29,000 killed
and who died of their wounds, and 16,000 from disease, at the cost of £220
million. Such costs were far from negligible.[1] With large numbers of Australians
involved in the conflict with many killed and wounded, the question arose as to
how to acknowledge their service and mark their sacrifice; a question answered in
part by a German soldier from a later war who wrote to his friends about how to
honour his memory. [He was killed the next day.]

> Farewell. You have known to all the others who have been dear to me, and you will
> say goodbye to them for me. And so, in imagination, I extinguish the lamp of my
> existence on the eve of this terrible battle. I cut myself out of the circle of which
> I have formed a beloved part. The gap which I leave must be closed; the human

WAR AND MEMORY IN AUSTRALIA

After the horrendous death tolls of the First World War, the making of public memorials was seen as the right way to honour the fallen soldiers, mainly because their families had been deprived of traditional mourning rituals, because their dead lay so far away. Similarly, 'Memory Keeping' of the dead after the Boer War, took various forms and included Honour Rolls, Avenues of Honour, obelisks or statutes. Once again different local communities and associations used a further variety of forms, such as the endowing of scholarships, specially dedicated buildings and significant purpose-built structures in public places such as bandstands and memorial gardens. Memorial tablets were placed in many local churches. As can be seen in the case of the fallen Caulfield Grammarians, 'memory keeping' took different forms. Apart from listing their names on school Rolls of Honour, memorial plaques were erected in churches or chapels for **Tom Stock, James Christie, Thomas Foster, Walter Spier** and **Stanley Reid. Walter Spier** was honoured with a fitting stone monument. **Ivor Wilkinson** was remembered in his local town of Maryborough by a suitably inscribed monument and grand water fountain in the middle of the town.

chain must be unbroken. I, who once formed a small link in it, bless it for eternity. And till your last days, remember me, I beg you with tender love. Honour my memory without gilding it, and cherish me in your loving, faithful hearts.[2]

Letter written by German soldier Otto Heinebach, the night before he was killed on 14 September 1916.

After World War 1, as in many other communities touched by the conflict, the making of Great War memorials (and their associated public ceremonies) in Australia, was a quest for the right way, materially and spiritually to honour the soldiers.[3] Historian Ken Inglis pointed out, that the families of the men killed in the Great War were deprived of the traditional mourning rituals of their culture, because their dead lay far away,[4] it not being feasible or financially possible to

BOER WAR MEDALS

Originally invented to recognise general military service in war after the Battle of Waterloo in 1815, British Army campaign medals were awarded to members of the armed forces for taking part in a campaign in a particular war. Awards for service in a particular battle or theatre within a war, often then took the shape of named clasps attached to the medal ribbons. Campaign medals are awarded in stark contrast to decorations of merit which recognise heroism or bravery. Two campaign medals were issued for the Boer War. The **Queen's South Africa (QSA)** medal is a silver and bronze disc with the crowned and veiled head of Queen Victoria on the obverse. The reverse side has Britannia with a flag in her left hand holding out a laurel wreath with the dates below of 1899–1900 to a party of advancing soldiers. In

the background are two warships and around the top the words SOUTH AFRICA. The ribbon has a broad orange central stripe flanked by two dark blue strips and red edges. Twenty-six different clasps were issued for the QSA. After Queen Victoria's death on 22 January 1901, when her son King Edward VII ascended the throne, the **King's South Africa (KSA)** medal was instituted. The **King's South Africa (KSA)** medal was awarded to all those who were serving in South Africa on or after 1 January 1902 and recipients also had to have completed 18 months service in the conflict prior to that date. The medal is silver with the bust of King Edward VII on the obverse with the reverse bring the same design as the QSA. The ribbon has stripes of green, white and orange and just two clasps were issued with the KSA.

transport the bodies of the dead soldiers back to Australia for burial by their families. A similar situation had arisen with Australia's involvement in the Boer War some 15 years before. In this circumstance, communities and regiments, for the first time in Australia's history, erected their own unique memorials to serve the purposes of mourning, commemoration and memory keeping, although the motivating reason was often made plain. As the plaque on the South African Soldier memorial in St Kilda Rd Melbourne put it, they were '... fighting for the unity of the empire which is our strength and common heritage.'5

The funeral of Lt. John Chrisp 5VMR who was killed in action at Vryheid in the Transvaal on 5 November 1901. Carrying the stretcher second from left is Lt Duncan Stock, the brother of deceased Grammarian Tom Stock. (AWM P11294.010)

Twenty-four Caulfield and four Malvern Grammarians served with various, Australian, British and South African forces during the 1899–1902 Boer War. Seven former Caulfield men paid the supreme sacrifice with their lives. How was their service and sacrifice acknowledged and remembered by their country, their communities and their former schools? How did CGS pay tribute to its former students who had not only served, but had sacrificed their lives in the effort during the Boer War?

In Canberra, various statues and tributes to recognise the service and sacrifice of Australians in war have been installed on Anzac Avenue adjacent to the Australian War Memorial since it was opened in 1965. It should be noted that it was not until 2017 that the National Boer War Memorial was unveiled to join the remaining memorials. The names of the seven Grammarians who died in the Boer War are commemorated on the bronze panels of the Roll of Honour at the heart of the Australian War Memorial.

It should always be remembered that the scale of the Australian enlistments for the Boer War were the largest in this country's history to that time. The loss of life was marked in significant ways, and especially so, given that no bodies were able to be repatriated home for burial in Australia. Each of the official colonial contingents has a war memorial that commemorates the names and memory of the service of fallen Victorians and other colonies. Not so for the 'private' regiments such as the Scottish Horse, which despite having lost 26 young Australians, and 20 Victorians, including Caulfield Grammarian Ivor Wilkinson, has no memorial anywhere in Australia.

In the words of the doomed German soldier, Otto Heinebach in WW1 just 16 years later, how did Australians remember their Boer War dead and honour their memory without gilding it? It is certain that the families of the lost Grammarians always cherished them in their loving, faithful hearts. How were the deaths of the seven fallen Grammarians marked and acknowledged?

Caulfield Grammar School had had nearly 700 boys pass through its doors during the years 1881–1899 and from 1899 to 1902 had enrolments in the low 100s. The school community, consequently, was close knit with headmasters, staff and fellow classmates knowing each student and their families. Losses in war were felt deeply, and the sacrifice of each fallen soldier was appropriately marked. It should also be acknowledged that in this time before mass communications as we know it existed, information about former students relied upon letters, the occasional newspaper article and word of mouth.

The first Caulfield Grammarian to fall in battle was Tom Stock who was killed in action on 9 February 1900. When news of his death was received at home a memorial service was held at Sandford just one month later. A suggestion was made during the service of a collection being made for a stained-glass window as a permanent memorial to Tom. The stained-glass window did not come to fruition, but in time a public subscription resulted in a finely worked brass plaque being installed in the church to Tom's memory.

At the conclusion of the Boer War local Shire Councils also erected their own memorials to the men of the district who had given their lives during the conflict. The Casterton Memorial with Tom Stock's name on it was inscribed:

Erected by the People of this District to the Memory of Those Brave Men of

Casterton Detachment of H Coy V.M.R who gave up their Lives for Queen and Country in South Africa. 1900.

In Hamilton, the major town of the district where H Company of the VMR had its headquarters, a fine monument was also erected with a similar inscription to that of Casterton. Again, Tom Stock's name figured prominently on the monument.

Caulfield Grammar School also recognised Tom's death, and a Melbourne newspaper reported on a special meeting at CGS one month after his death.

A gathering of the Old Boys of the Caulfield Grammar School assembled in the school hall on the evening of 14 March, to make arrangements for the erection of a memorial tablet in the main classroom to the memory of Mr Tom Stock, a boarder at the school from 1892 to 1896, who fell in battle at Rensburg, gallantly fighting for his country. A resolution was passed, directing the secretary [of the Old Boys Association] to forward a letter of sympathy to Mr Stock's brothers and sisters at Sandford on behalf of the Old Boys.[6]

Following this meeting, Walter Murray Buntine, the CGS Headmaster, also wrote to the Stock family.

Please allow me as Tom's old school master to express my sincere sympathy, as well as that of our school, for you in the loss you have sustained through his death in South Africa. We are proud to own him as an Old Boy of the Caulfield Grammar School and to think of his brave and noble deed. Your loss is our loss and our country's. I knew Tom sufficiently long enough to be aware of the splendid qualities he possessed and was not surprised to read of him as I have done lately, in the columns of the daily papers. Amid the trial you are called to bear, it must be a great consolation to remember how nobly he has served his country. The Old Boys of this school have, I believe, already in hand a considerable sum to erect a tablet to his memory in the Dining Hall or the main Schoolroom. With sincere sympathy, I am W.M. Buntine.[7]

The subscription for funds was most successful, and a striking memorial tablet was unveiled at CGS in the main hall just six months after his death and the entire ceremony was reported in Tom's local newspaper the *Casterton News*.

Memorial tablet to the late Private Tom Stock
unveiled at Caulfield Grammar School

A tablet to the memory of the late Mr Tom Stock, who was educated at Caulfield

Grammar School during 1893–1896, has been recently erected in the main hall of the school by a few of his old school fellows; and ceremony of unveiling took place on the evening of Friday last, before a large gathering of past and present pupils and their friends. The chair was taken by Mr MacKenzie, MLA and speeches were made by the chairman, by Major Reay, by several Old Boys and by the Headmaster Mr WM Buntine.

The headmaster said that it seemed almost impossible that one of the boys of the school who would be best remembered by his prominence in all school concerns, his cheerful nature and his boyish love of sport, in which he distinguished himself in contests with school teams, could, in little more than three years, have found a soldier's grave on the African veldt. Men appear as they are, in a crisis. Great difficulties call forth for the greatest effort and he had scarcely known Tom Stock, for the fact that the great occasion was wanting to show in his true light. He was always an earnest boy possessed by a strong determination. These qualities were easily detected in studies and sport, and no doubt went far in him and in his comrades to enable them to write up an undying reputation on the first occasion when Victorian soldiers were put to the test of war. He was too, a generous boy who was ever ready to make allowance for the faults of others and had won for himself a very strong attachment by his school fellows. The Old Boys had done a wise thing in perpetuating his memory in the way they did, for boys are natural hero worshippers, and the boys of Caulfield Grammar School would have in Tom Stock a worthy hero.

Mr MacKenzie, the chairman, said that he could not speak but with mingled feelings of pride, pleasure and pain. It was a matter of pride for the Caulfield Grammar School and for the whole colony. Tom Stock and his gallant companions had upheld the best traditions of their race and had acquitted themselves so nobly in the recent campaign. It would have been a pleasure had it been possible to welcome them back to our shores, but that was impossible. They went prepared to give a sacrifice, if necessary, for the cause of the Empire and we knew that Australians would acquit themselves as well as the race from which they had come. As in Roman History, what Rome held most dear, her bravest son, had on one occasion to be given up to death that the nation might be saved; so, our best lives had been given to prove our loyalty to the great British Nation. A tablet such as that erected would be the link to bind us to the motherland. No union was so strong as the cement of blood spilt and mingled together on the veldt of South Africa.

Major W. T. Reay (The War Correspondent for the *Melbourne Herald* newspaper) very feelingly referred to his acquaintance with the father of the late Tom Stock and spoke of the esteem in which he was held in the Western District. Tom had been a member of his own regiment and had shown himself possessed of the fine qualities of a fine soldier. He (Reay) had gone to South Africa as a war correspondent and was with Tom when he was killed in the fight at Pink Hill, where that day, in the death of Stock he lost a dear friend and comrade. A graphic account was then given of the situation of the Boers positions and the way in which they drove in the British outpost on that fatal day.

Short speeches were then given by Mr T.V. Sewell, who had acted as secretary for the fund towards erecting the tablet and by another Old Boy of the School, Mr A.M. Lonie, of the firm of Gaunson and Lonie, solicitors.

The tablet, which was covered with an Australian flag, being unveiled, the proceedings were ended by the production of some interesting war views by Mr Alex Gunn.

The tablet read: Erected by his school fellows to the memory of Tom Stock a member of the first Victorian Contingent and a former boarder of this school 1893–1896. Who was killed in battle while fighting for Queen and Country near Rensburg, South Africa, February 9th A.D. 1900. Labora Ut Requiescas.[8]

For many years the Tom Stock plaque hung in the hallway near the old staffroom and next to the Honour Rolls from World War 1 and World War 2. With the construction of the War Memorial Hall in 1958, the plaque was relocated to its foyer but was taken down in the rebuilding of the Halls and since 2000 has been safely housed in the CGS Archives.

Caulfield Grammarian Thomas Foster had succumbed to enteric fever on 22 August 1900 and was buried at Umtali in Rhodesia. In addition to his name being recorded on the CGS Roll of Honour, and the Inglewood War Memorial, Ballarat College, his other old school also marked his passing.

A memorial tablet erected to the memory of Ballarat Collegians, who fell in the South African war, was unveiled by the principal Mr John Garbutt MA, at the college on Wednesday 8th April 1903. The tablet, which was provided by the Old Boys' Committee, bears the following inscription: 'In memory of Old Collegians, who died in the war in South Africa, 1899–1902. Private T. B. Foster. [and the names of three others] Their duty done, they rest with God,[9]

The second Caulfield and Melbourne Grammarian to die as result of contracting enteric fever was James Christie on 7 December 1900. He was formally remembered in three memorials, the first of which was noted in a school magazine

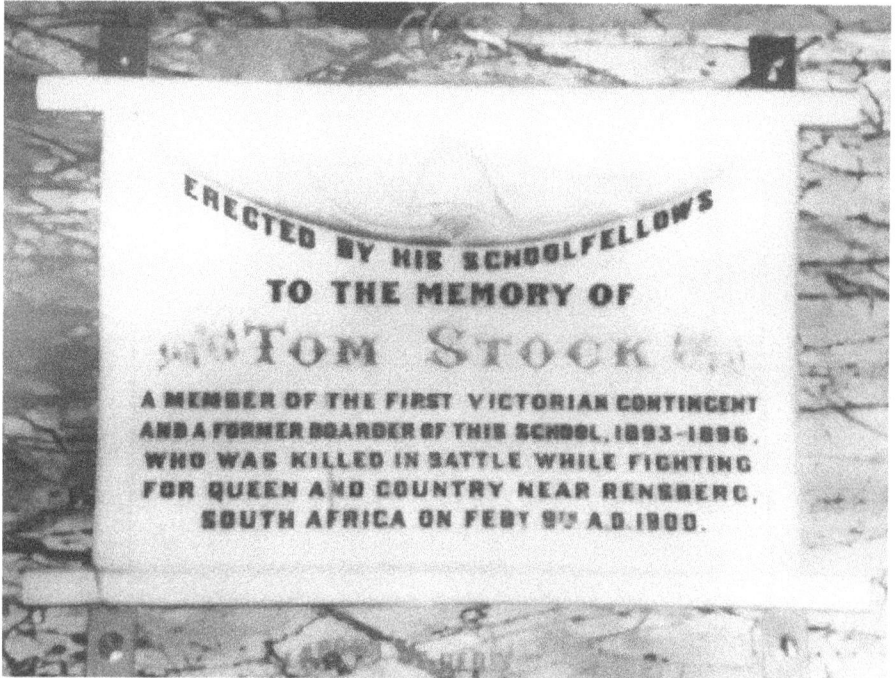

The commemorative plaque at CGS in memory of Tom Stock, the first Caulfield Grammarian to be killed in war. (Author's collection.)

article by Melbourne Grammar School.

> The Rev. A Caffin, Vicar of St Matthew's Cheltenham, solicits subscriptions to a memorial window to be erected in his parish church to the memory of J.W. Christie, an Old Melburnian who entered the school in 1884. [incorrect date] Christie was the only son of a widowed mother and was a surveyor under the Metropolitan Board of Works. He was a private in the Bushmen's Contingent and died at Rustenburg on 30 November last year. The amount required is £20 and the local subscriptions to date are £11/3/6.[10]

But the plans appear to have been changed and a news article in July 1901 noted that.

> A most interesting military service took place last Sunday afternoon at St Matthew's Church, Cheltenham, when a memorial tablet was unveiled to Pte. Christie of the Bushmen's Contingent. The tablet is carved out of solid marble and is a beautiful piece of work was unveiled by Colonel Tom Price and supported by several of the G Company Rangers. The Headquarters Band played the Dead March, and the church choir led the singing. An overflow meeting was held in the

school room which was addressed by the gallant Colonel who spoke most feelingly [about Christie] to the children and young men present.[11]

Another newspaper article published in 1904, outlines in some detail a second memorial which was unveiled in 1901 in Cheltenham which mentions James Christie. Under the heading, 'Cheltenham Memorial to the Moorabbin Volunteers Unveiled,' the article read in part.

> On one side are inscribed the words – 'Erected by the residents of Moorabbin to perpetuate the noble deeds of our Moorabbin boys whose names are inscribed volunteered and fought for the Empire in the South African war, 1899–1902.' On another side the record is made – 'Unveiled by the Hon. T Bent, Premier on May 7th1901.' High up on the column are the words: 'In Memoriam – Trooper W. Christie, died at Rustenburg, 1900.[12]

Finally, Melbourne Grammar School marked the Boer War service and deaths of some of its former students in the following manner.

> The windows which have been placed in the chapel of the Melbourne Grammar School in memory of the late Professor E F Morris, headmaster from 1875–1882, and of the Old Melburnians who fell in the Boer War were dedicated yesterday by Archbishop Clarke. The following old boys of the Melbourne Grammar School were either killed in action or died in South Africa during the Boer War. Messrs J W Christie ...' [and another seven names].

The third former Caulfield and Sydney Grammarian who succumbed to enteric fever, was Walter Laishley Spier who died at the Woodstock Hospital in Capetown on 23 January 1901. He is remembered on memorials at Hunters Hill and at Sydney Grammar School whose magazine stated when their plaque was erected in 1903 that it was done:

> As a memento of those whose untimely death they were mourning and was a concrete testimony of the patriotism, loyalty and self-sacrifice of those who had fallen in what they believed to be a righteous cause.[13]

Another memorial was also under consideration as reported in an article headed 'The Late Corporal Spier' in a 1901 Sydney newspaper.

> At a well-attended meeting of the residents of Hunter's Hill and surrounding districts held at the council chambers, Hunter's Hill, on Tuesday last, at which Mt Henry Deane, in the absence of the Mayor, occupied the chair, it was unanimously decided on the motion of Sir George Dibbs, seconded by Mr Justice

Memorial to the Late Corporal W.L. Spier

A.H. Simpson, to erect a monument to the memory of the late Corporal Walter Laishley Spier, of the Bushmen's Contingent, who died at Cape Town on January 23 last. An influential committee was formed to take the necessary steps. Mr Lawrence Stephenson was appointed Hon. Treasurer and Mr J.J. d'Apice the Hon. Secretary to the fund.

Eventually the monument was erected in Hunter's Hill and dedicated on Sunday 26th May 1901, an event reported in the Sydney newspapers.

It is a polished red granite cross with a base on a larger rough-cut stone set on a concrete slab within a low iron railing, located within the grounds of All Saints Anglican Church, Hunters Hill. The formal unveiling of a memorial to the late Corporal W. L. Spier, of the New South Wales Bushmen's Contingent, who died of enteric fever in the Woodstock Hospital, Cape town, on January 23 last, was performed on the 26th ultimo by Captain Sir George Dibbs at All Saints' Church, Hunter's Hill. The monument is of artistic design and is placed in a prominent position at the western end of the church grounds. It is of polished granite and was the work of Messrs. H. Taylor and Sons, monumental masons, of Sydney. The ceremony took the form of a church parade, an imposing military force being present, composed of about 100 men of the National Guard under the command of Sir George Dibbs (captain commanding), who was assisted by Captain W. Henderson and Lieutenants McDonald and D. Inglis, a large detachment of the returned Bushmen's Contingent, Mounted Rifles, and Australian Horse, under the command of Lieutenant W. Cope; and representatives of the Artillery and other regiments.[14]

Former Caulfield Grammarian Stanley Reid had spent most of his school days at Scotch College but was noted at the 1901 CGS Speech Night, after he died of wounds on 23 June 1901. After hostilities ceased in May 1902, Reid's body, along with others was reburied at Middelburg Cemetery and he is commemorated on memorials at the Australian National War Memorial in Canberra, Caulfield Grammar School, Scotch College in Hawthorn, Ormond College at Melbourne University, and in King's Park, Perth Western Australia.

The last Caulfield Grammarian to be killed in action in the Boer War was Ivor Wilkinson who lost his life on 3 July 1901. In a letter written to the author in 1992, Ivor's great nephew Peter Ross, remarked that he had lived all his life with his grandmother Maude, one of Ivor's sisters, who always spoke in such glowing terms of Ivor her lost soldier brother, that no matter what Peter did, he could never aspire to be the paragon of virtue as Ivor was portrayed. Mr Ross also expressed the opinion that he personally had a great sorrow at the loss of what was obviously a fine young man with a good sense of humour, a love of sports and a zest for adventure. He hoped that future generations would keep in mind that Ivor lived and died in very different times, and that in the terms of those days he was doing what was expected of him and the important thing was that he had done it well.

> Perhaps no more and no less than many others, but for those of us in his family and those from his school, it's not amiss to feel a little pride.[15]

The town of Maryborough also sought to express its pride and grief at the loss of their first citizen in war. On 22nd April 1903, after a most successful public subscription for funds and in the presence of 1300 spectators, the local MP unveiled a handsome drinking fountain in front of the Town Hall in memory of Gerald Massey Ivor Wilkinson, the sixth Caulfield Grammarian to die during the Boer War. Wilkinson's name is also registered on the CGS Roll of Honour.

Although he was mentioned by Principal Buntine at the 1900 Speech Night and lauded as a member of the 1st Victorian Contingent, alongside Tom Stock, Sergeant Major Ernest Norman Coffey, the CGS Cadet Drill Sergeant, received no mention by CGS again. Perhaps it was, that when he returned from active service in South Africa, he was unable to resume his duties due to illness, and that he passed out of the attention of the school. Nonetheless, his works for CGS and his service and subsequent sacrifice deserves to be acknowledged, if only to

A grave on the South African Veldt at Utrecht, Natal for Farrier Sergeant R. Albert Houghton a member of A Company, 5th Contingent of the Victorian Mounted Rifles who was killed in action on 16 January 1901. A column of mounted troops is on the move in the background. (AWM P01866.005)

recognise the ongoing injustice accorded to his wife and family. The Boer War was declared officially over on 31 May 1902. Unfortunately for Coffey, the good wishes extended to him at Richmond, did not prevent him from succumbing to his ill-health and the condition he had contracted in South Africa. He died of tuberculosis at his Wellington Street home in Box Hill on 18 September 1902 as a newspaper of the time noted.

> Many military comrades and friends will learn with very deep regret of the death of Regimental Sergeant-Major Coffey of the 2nd Battalion Infantry Brigade. He died from phthisis, the result of a cold contracted in South Africa, whither he went with the First Contingent, under the late Major Eddy and served for twelve months. During this campaign the Sergeant-Major proved himself a most capable and zealous man, and his valuable services when the infantry converted into mounted troops will be especially remembered. The deceased will be accorded a military funeral tomorrow afternoon, starting from Christ Church Hawthorn. It is notified elsewhere that members of the 2nd Battalion who wish to attend [the funeral] may do their firing practice in the company and battalion firing practice early tomorrow, commencing at noon.[16]

Incredibly, upon his death, his widow who was supporting four children was denied a pension, because it was deemed that Coffey had not been killed in action or died during the duration of the war!

A letter from Mr Morgan P Kenny, the [Victorian] President of the South African Soldiers' Association was published in the AGE in January 1920 to highlight the injustice.

> There are many cases [of injustice] within our midst which require explaining by the State Treasurer. Take the case of former Sergeant-Major Coffey, who died from disease contracted whilst on active service with the first Victorian contingent to the Boer War. As Sergeant-Major Coffey did not die until eight days after the time allowed by the act, his widow was deprived of any pension. Instead, after representation being made, she was granted a compassionate allowance which was stopped in June 1915, without any reason. Is this the way the State Treasurer honours the promise of the State Government, who told the soldiers when they returned from South Africa – 'A grateful country, will never forget the services and sacrifices you have made on behalf of the Colony of Victoria.[17]

A legal battle had been waged, mainly by his former commanding officer's law firm, to try and ensure that his widow received an appropriate pension and justice. A 1920 newspaper article outlined the details.

CASE OF MRS. COFFEY. [LETTER] TO THE EDITOR OF THE AGE.

Sir, In the report in your issue of today the State Treasurer is reported to have said, 'As far as he knows there was no widow of a South African soldier receiving a pension from the State.' The Treasurer is in error. Regimental-Sergeant Major COFFEY contracted tubercular disease on service, which resulted in his death. During his Lifetime a pension of 3/0 per day was paid to him, in accordance with the South Africa Contingent's Pensions Act No. 1997 of 1905. On behalf of Mrs COFFEY, we made representations to the Treasurer and pointed out to him that by the act quoted pensions were granted to the widows and dependants of soldiers killed on service and that we thought Mrs. COFFEY should receive the same consideration as was given to the dependants of soldiers actually killed in action in the South African War. Not to have granted a pension to his widow was, it appeared to us, either an error or an outrage. As we assume it was an error on the part of the Government in 1905 in not granting a pension to Regimental-Sergeant-Major COFFEY's widow, and we so informed the Treasurer by letter on 5th October 1918. The Treasurer, on 17th October 1918, replied that as Mrs. COFFEY was eligible for an old age pension, she should apply for same, and he pointed out that the cases of pensions granted to widows by that act were 'quite different from the case of Mrs COFFEY, as their husbands were killed in

action.' On the 21st October, 1918, we replied to the Treasurer pointing out
there was no substantial difference, as Regimental-Sergeant-Major COFFEY had
been through the actions in which the other soldiers had been killed; and had
rendered good service before contracting on service the disease which resulted in
his death, 'and therefore Regimental-Sergeant-Major COFFEY gave his life in his
country's service just as did the men who were killed in the actions in which he
had fought.' Our interest in the matter as not merely professional. It is personal,
as our Lieutenant-Colonel MCINERNEY was the officer commanding the unit
in the First Contingent for service in South Africa in which the Late Regimental-
Sergeant-Major COFFEY served in the South African war, and he knows how
faithfully the Late Regimental-Sergeant-Major COFFEY discharged his duties on
service. Yours &c., MCINERNEY, MCINERNEY and WILLIAMS. 27 January
[1920].[18]

Although various Patriotic Funds had granted Mrs Coffey some small amounts
of money in the years after his death, Coffey's case seemed to have fallen into the
'too hard basket.' He had been sent to war in 1899 by the Victorian Government
but had returned to a Federated Australia. His widow and family had then to
make appeals to both an apparently uninterested Australian Government, which
by then was controlling military matters.

Apart from these tangible measures taken in various communities and some
in schools, in the days before emails and the internet, CGS tried to keep its
community informed about the involvement of its former pupils at the Boer
War at its large gatherings. The 1900 CGS Speech Night program made the
first mention of the war and tried to outline the involvement and sacrifice of
Grammarians. At this stage of the war CGS appears to have no knowledge of
other Grammarians who had enlisted in other colonies' military forces.

OLD BOYS AT THE WAR

Several members of the School Cadet Corps of past years took active part in the
recent war, where they appear to have displayed the courage and gallantry of
veterans. With the First Victorian Contingent there were Private Tom Stock and
the School Drill Instructor, Sergeant Major Coffey. With the Australian Imperial
Regiment [4VIB] there were Privates R. Holloway, A. Anderson and T.B. Foster.
Lieutenant George Macartney, son of Rev.H.B. Macartney, who was with the 2nd
Royal Fusiliers, and who was dangerously wounded in the storming of Pieter's Hill,
was also many years at Caulfield Grammar School. At Pink Hill, near Rensburg,
whilst making a stubborn defence in company with several other gallant fellows,

Private Tom Stock fell mortally wounded amongst the first Australians to die for the Empire. His school fellows, as a token of their admiration for him, and to perpetuate the memory of his bravery, have fixed a marble tablet [August 1900] in the large school room.[19]

This marble tablet became the school's first memorial of any description until the erection of the Roll of Honour in 1919 at the conclusion of World War 1. Badly wounded Grammarian Reg Holloway's visit to CGS and other Boer War news was announced in the 1901 Speech Night program.

OLD BOYS AT THE WAR

Since the death of Mr Tom Stock at the beginning of last year, which took place in the first engagement where Victorians took an active part, and the wounding of Lieutenant George Macartney in the relief of Ladysmith, many other old boys of the school have volunteered and gone out to South Africa to fight for the honour of the flag and Australia. Amongst those who have given up their lives nobly fighting for the Empire, the boys of our school who have gone to the war seem to take a somewhat disproportionate part. Private T.B Foster [4VMR] died of enteric fever in a field hospital after being present in several engagements. Lance-Corporal Wilkinson died of wounds received in action while connected with Tullibardine's Scottish Horse. Only one of out eight has so far returned, and he after being very severely wounded. The return of this soldier, Private Holloway [4VMR], was enthusiastically celebrated at the school some months ago. There are still in the field Sergeant Birch [5VMR] and Privates McPherson [5VMR] and Anderson [4VMR], from whom several interesting letters have been received. Lieutenant Stanley Reid, who volunteered as a Chaplain and was afterwards accepted as a Lieutenant in the fourth contingent [actually 6WAMI] and was killed in action during the earlier part of the year, was for some time [one term] a boarder at the school. We anxiously hope that the remaining three representatives of the school will be fortunate enough to return to us safely ere long.[20]

This last sentence indicates a lack of contact and communication between the school and its Grammarian Boer War soldiers. At least twenty-three former CGS men are known to have enlisted for service during the Boer War. At least seven soldiers included in this number enlisted in the contingents of other states and so a lack of knowledge by CGS of the military service of these former students could be quite understandable. But at least fourteen Grammarians were members of Victorian units, yet not all of them, are ever mentioned in CGS publications or records. This silence about the remaining Grammarian Boer War veterans is further emphasised in the early days of the Great War some 13 years later. The June 1915 issue of the CGS School Magazine in the context of constructing a Roll of Honour for World War 1 servicemen mentions the Boer War service of Grammarians, but still omits many names.

THE BOER WAR

When then Roll of Honour is finally placed in the school, it is intended that the completed list of those who were on active service with the Australian contingents in the Boer War, shall be given a place beside those of their school fellows who

AUSTRALIAN BOER WAR MEMORIAL

On 31 May 2017, the Governor General Sir Peter Cosgrove dedicated a memorial to all those Australians who had volunteered, fought and died in the Boer War. In 2009 a National Boer War Memorial Association (NBWMA) had been formed and a data base established of descendants of Boer War soldiers. In time the NBWMA produced a quarterly newsletter, a site was selected for the memorial and a design brief completed. An international design competition begun in 2009 and took two years to complete before the final design was approved by all parties. The Australian Boer War Memorial is located on Anzac Parade in Canberra alongside Australia's other significant war memorials and can be described as follows:

'The Memorial represents a half section (four troopers) of mounted riflemen on patrol in South Africa.

The patrol is shown in motion and their horses step carefully through the sparse and arid landscape sloping down from the back of the memorial. Each soldier keeps an eye on his mates while looking out for signs of booby traps or the potential for an ambush or for the tell-tale wheel tracks of big guns or supply wagons. Or for snipers or smoke from campfires. Friend or foe? Their dress is patterned after that worn by the Australian Commonwealth Horse (ACH) regiments of 1902, Australia's first official national units. These mounted men had their gear set up so that if they parted company with their horse, the soldier took with him his weapon, ammunition, water bottle and his emergency rations.'

https://www.bwm.org.au/memorial_story. php

have enlisted with the Australian Expeditionary Forces. It is fitting that we should think of them at this time, and we give here their names.

Casualties: Thomas B Foster, died of enteric. Reginald J Holloway severely wounded. H.B. George Macartney severely wounded. Stanley S Reid died of enteric. Tom Stock, killed in action. Ivor Wilkinson, died of enteric. [killed in action]

Others: Frank V Weir, mentioned in despatches for gallantry. A. P Anderson.

Once again, at the time, a lack of knowledge from CGS about its Boer War veterans is clear. For example, two casualties mentioned, did not succumb to enteric at all, as Stanley Reid was mortally wounded in battle and Ivor Wilkinson was killed in action. In addition, to the lack of mention in any CGS publication of at least fifteen other Grammarian Boer War veterans, the school makes no mention of three other deaths. Two of these were former students, James Christie (1885–86) and Walter Spier (1888) who had attended CGS, but had then transferred to Melbourne Grammar and Sydney Grammar respectively. Along with Ernest Coffey, their names have never appeared on any CGS Roll of Honour until 2025. This is stark contrast to Stanley Reid, who attended CGS for just one term in 1886 before transferring to Scotch College and spending some seven years there. He has been 'claimed' by both schools and his name has been listed on the CGS Roll of Honour since his death.

Following the end of the Boer War, CGS did not install a permanent record or Honour Roll of the services of its veterans. The matter of the installation of a Roll of Honour at CGS, was not mentioned again in any record until the School Magazine of December 1919 and Brigadier General Brand made mention of the Great War military service of Grammarians at its unveiling.

THE UNVEILING of the HONOUR ROLL

The Honour Roll was placed in the hall, close to the main door and was draped with a Union Jack. Many had fallen on the battlefield; these deserved to be honoured even more than the winners of the VC and to those who had lost relatives and friends, he extended his sympathy. The Honour Roll is now in the School Hall as a permanent mark of what their predecessors did in the Great War. It contains the names of approximately 440 old boys, of whom 66 laid down their lives.[21]

Despite earnest expressions of intentions expressed by CGS in June 1915 concerning the listing of the names of the Boer War soldiers, their names were omitted from this first ever CGS Roll of Honour. This oversight was not rectified until the construction of the Caulfield Grammar School War Memorial Hall in 1958 when the names of Tom Stock, Thomas Foster, Stanley Reid and Ivor Wilkinson were added to a newly commissioned Roll of Honour. The names of James Christie, Ernest Coffey and Walter Spier were not included.

Contemporary Australian scholars such as Professor David Horner, saw that the conflict in South Africa helped to prepare the Australian military forces for future conflict. Although the conditions under which the Boer War was fought essentially involved mounted troops, and the Great War often bogged down into one of dreadful trench warfare, Horner noted that:

> The Boer War showed the value of training, discipline, leadership and administration to anyone who was willing to learn. An appreciation of these timeless military qualities was invaluable in the newly raised Australian Imperial Force (AIF) in World War 1. It is likely, then, that the experience of the Boer War had a greater influence on the commanders of the 1st AIF,[22] particularly in the early stage of World War 1, than has been generally recognised.[23]

Twenty-four Caulfield Grammarians and four Malvern Grammarians are known to have served in the Boer War in South Africa between October 1899 and May 1902. Seven lost their lives. What became of the CGS and MGS Boer War veterans after their service in South Africa was completed? Of the remaining twenty-one men for example, when the First World War was declared in 1914, nine Caulfield and three Malvern Grammarians, twelve men in all, enlisted for military service again. However, the lives of all Caulfield and Malvern Grammarian Boer War survivors are most compelling to read about.

Endnotes

1 Hobshaw. E.J. *The Age of Empire 1875–1914*. Weidenfeld and Nicolson Ltd. 1987. p306.

2 Hanson N. *The Unknown Soldier. The story of the missing of the Great War*. Doubleday. London. 2007. Preface.

3 Inglis B.S. *Sacred Places*. ibid. p122

4 Inglis B.S. ibid. p93

5 Hutchison G. *Remember Them. A Guide to Victoria's Wartime Heritage*. Hardie Grant Books. Prahran. 2009. p19.

6 *Weekly Times* (Melbourne) Saturday 24 March 1900 p15.

7 Wilmot. P. ibid. p18.

8 *Casterton News*. Thursday 2 August 1900.

9 *Argus*. 9 April 1903. p6

10 Melbourne Grammar School magazine. *The Melburnian*. 1901. No. 26. p15

11 *Brighton Southern Cross*. 13 July 1901. p2

12 *Mornington Standard*. 21 May 1904. p3

13 Sydney Grammar School magazine – 'Sydneians in the Boer War' SGS Winter 2016. p39

14 *Sydney Mail and New South Wales Advertiser* (NSW), 15 June 1901. p1484

15 Ross P. M. Letter to D.J. Moran. 13 December 1992 (Held in the CGS Archives)

16 *Herald*. Friday 19 September 1900. p2

17 *Age* Thursday 22 January 1920. p6

18 *Age* Friday 30 January 1920. p6

19 CGS Speech Day program. December 1900. p12.

20 CGS Speech Day Program. 17 December 1901. p11.

21 CGS Magazine. December 1919. pp4–5.

22 David Horner notes that of the twenty notable generals who led Australian forces in World War 1, fifteen had served in the Anglo-Boer War and that of the seventy-six brigadiers who had direct involvement in leading Australians, thirty-three had also served in the conflict in South Africa.

23 Horner. D. *Strategy and Command – Issues in Australia's Twentieth-Century Wars*. Cambridge University Press. 2021. p23.

CHAPTER 9

THE FINAL RECKONING FOR BOER WAR GRAMMARIAN VETERANS

PREAMBLE

Of the Caulfield and Malvern Grammarians who served in and survived the Boer War, returning home was never an easy task. Some received a grand welcome, some married, some dealt with ongoing debilitating wounds, while others quietly slipped back into their former lives and faded into obscurity, and we have little trace of them. Just twelve or so years later, after their Boer War service, nine of them enlisted for military service in the Great War and sadly, four of them lost their lives. The short profiles that follow are an attempt to learn more about the lives of these Grammarians following their Boer War service between 1899–1902.

Andrew Percival ANDERSON (1884–89?)

Anderson served with the 4VIB until 22nd June 1901 and after his service was completed, he was granted a passage to England. At the conclusion of hostilities in 1902 he was entitled to the QSA with four clasps. Eventually he returned to Australia and married Frances McDonald at Prahran in 1905 passing away at Brighton in 1948.[1]

Bernard Everett BARDWELL (1892–96?)

Bernard and his younger brother Beresford both moved to Broome in West Australia after the Boer War and purchased two luggers and established a successful business in the pearling industry. On the outbreak of the First World War, Bernard enlisted as a Lieutenant in the 16th Battalion and served on the Western Front before being promoted to Captain in the 48th Battalion. After being badly wounded in the leg in July 1916, he was later admitted to hospital in November with heart disease and was eventually repatriated to Australia, being discharged in July 1917. After the war, he married Kathleen Sefton, had two children and then resumed his operations in the pearling business. In 1920 the brothers landed a large pearl worth about £4000 or $700,000 in today's value. Whilst his brother Beresford became the Broome Harbour Master, Bernard

Lt Bernard Bardwell pictured in Egypt en route to the Western Front in WW1.
(https://broomeanzacs.broomemuseum.org.au/?page_id=160)

became a Fisheries Inspector, and in his spare time became a collector of shells and was recognised as one of the world's outstanding conchologists. Bernard passed away in 1955.[2]

Ormonde Winstanley BIRCH (1896–98)

Upon his return to Australia after his service in South Africa, Ormonde Birch was selected as one of the few returned men to be a member of the Edward VII Coronation Contingent. British authorities had invited Australia to send a contingent of troops to the Coronation of King Edward VII in Westminster Abbey on 26th June 1902. The Australian Government agreed to send mounted troops chosen from those who had exceptional service in the war in South Africa or were still serving in South Africa. The contingent comprised about 150 men from all states, who wore specially designed uniforms, and departed from Sydney

A KING'S CORONATION AND AN EMPEROR'S DURBAR

A KING'S CORONATION.

Edward VII and his wife Alexandria as the King and Queen of the United Kingdom and the British Dominions, held their Coronation in Westminster Abbey, London, on 9 August 1902. Originally planned for 26 June 1902 it had been postponed as the king had required emergency abdominal surgery to treat appendicitis. As a result of the postponement, many of the foreign dignitaries had had to return home. Consequently, many countries were represented just by their ambassadors, among the 8,000 other guests. The Coronation Procession of royal carriages was flanked by 2000 troops and included members of the British Army, Royal Navy, Royal Marines and Colonial, Indian and Dominion forces. Amongst this last group was **Ormond Birch** who, with all other soldiers, received a bronze King Edward VII Coronation medal. Remarkably, on 16 August 1902, the King held an

audience aboard the royal yacht where he welcomed three of the leading Boer commanders, Louis Botha, Christiaan de Wet and Koos de la Rey, just a few months after the Treaty of Vereeniging had brought the Boer War to an end.

AN EMPEROR'S DURBAR.

The Delhi Durbar (Court of Delhi) on 12 December 1911 was designed to allow the King and Queen to be proclaimed as the Emperor and Empress of India and was the only time that British monarchs ever attended the Durbar in person. Over 40,000 British troops paraded at the event in front of 250,000 spectators. The event was mainly designed to cement support for British rule among the Indian princes and people.

Ormonde Birch was also present and officiated in a church service, after which he was presented to the King who spoke to Birch about the Boer War and Coronation medals, he was wearing.

Six members of the 1902 Australian Coronation Corps on board ship for return to Australia having participated in the Coronation Parade for King Edward VII. Each man wears his medal for service in South Africa as well as the Coronation Medal which they all received. (AWM A04831)

on the passenger ship *Rome,* on 30th April 1902. Australian Prime Minister, Edmund Barton and his wife were also on board with many other civilian passengers. Some of Birch's observations and experiences in London as part of the Coronation Contingent were published in a Victorian newspaper in 1902.

We are having a splendid time and are being treated right royally and the [Coronation] service at St Paul's was very impressive. We have received no end of invitation to theatres, concerts and private houses and I have not had time to go to half the places at all. We had a most interesting visit to the Woolwich Arsenal to see guns and cartridges being manufactured. On 9 July we went to Windsor Castle and had a fine reception, where the toast was 'The King and Royal Family and the loveliest Queen who ever sat on the throne of Great Britain.[3]

On his return from this special duty, he studied at St Wilfred's Theological College in Cressy, Tasmania, being appointed a deacon in June 1906 and was ordained as a priest a year later. From 1907–1910 he worked at Holy Trinity in Hobart, before spending three years in England serving as the acting military chaplain at

Aldershot in 1910–11.

He moved to India in late 1911 as the probationary chaplain to the Bengal Establishment and later served as the Chaplain at the Ecclesiastical Establishment in Dinapur between 1911 and 1914.[4] He attended what was billed as the greatest event of the British Raj in India, when in December 1911, King George V attended the Delhi Durbar, becoming the only reigning British monarch to ever do so. Over 50,000 British soldiers took part in the parade and associated festivities and a special medal was awarded to all the participating troops, including Birch. A 1914 newspaper article explained more about the Rev. Ormonde Birch and his involvement in this event.

> Rev. Birch is at present filling the office of military chaplain and honorary member of staff at Dinapur, India to the troops assembled there. He is a keen cricketer and played for North Hobart in the district competition. At the Delhi Durbar, King George went to Arrah to inspect a parade of troops. Mr Birch, together with the bishop and his chaplain, held a parade service and was afterwards presented to the King, who noticed his Coronation and Boer War medals and asked him many questions regarding his career and Australia.[5]

When the First World War was declared he served with the British Expeditionary Forces as a chaplain in Mesopotamia between 1915 and 1918, twice being mentioned in despatches. He was promoted to Chaplain Major and in August 1917 was awarded the Military Cross for *'untiring devotion to duty.'* At war's end he returned to Australia and married Amy Frances Shoobridge in Hobart in August 1921. The couple returned to India the following month and he served at Kasauli, Patna, Calcutta, Shillong and Kidderpore between 1921 and 1932. In 1932 Ormonde was appointed to St John's in Calcutta where he served as Archdeacon until 1935, before being appointed as the Chaplain at the British Embassy in Madrid in 1936.

He returned to Australia and served the Anglican Church in appointments in

Chaplain-Major O. W. Birch, M.C.

Ormonde Winstanley Birch (https://www.caulfieldgrammarians.com.au/alumni-profiles/ormonde-winstanley-birch/

GRAMMARIAN GALLANTRY WAR MEDALS AND AWARDS

The British Armed Forces recognised the bravery and gallantry of its members by the awards of various awards and decorations. As Australia had no such honours system in place during the Boer War or the World War I, these awards were also made to Australians including some Grammarians as noted elsewhere in the text. The **Distinguished Service Order** (DSO) was instituted in 1886, and is an award made to officers for 'Meritorious and Distinguished services during active operations against the enemy usually in combat conditions.' The **Military Cross** (MC) was instituted in 1914 and was awarded to officers who had, 'Displayed acts of exemplary gallantry during active operations against the enemy.' The **Military Medal** (MM) was first established in 1916 and was awarded to military personnel below commissioned rank (officers) for bravery in battle on land. Being **Mentioned in Despatches** (MiD) meant that the person concerned had performed some act or service that warranted their name being included in an official account of the campaign or action. Awardees received a certificate and were entitled to wear an oak leaf decoration on their medals.

Kew, Brisbane, Melbourne, Armadale and Caulfield in the period 1937–1946. Elevated as a Venerable Archdeacon, Birch lived out his remaining years in Malvern, passing away there in 1969.[6]

William BOYD (1886–90?)
Little is known about William after his Boer War service, but he appears to have returned to the Inglewood area.

James William CAMPBELL (1884–91?)
After the Boer War, James (Jim) Campbell become a copra planter in Bougainville and established the Soraken Plantation for Burns Philp and Co in the years before the First World War.[7] Campbell was also for a short time an Inspector for the Copra Board and operated several trading stations along the coastline near the Buka Passage in Bougainville.[8] On the outbreak of the First World War, he enlisted (3046) in the 3rd Light Horse AIF on 15 May 1916 and served in Palestine being promoted to sergeant. For some reason he relinquished this rank and transferred to the infantry in the 32nd Battalion and was posted to France. On 8 August 1918 the Allies launched a major offensive at the Battle of Amiens

in which the 32nd Battalion participated. An Australian War Memorial article provides the context for the award of a gallantry medal to James Campbell after this action.

> [The 32nd Battalion] was subsequently involved in the operations that continued to press the retreating Germans through August and September. The 32nd fought its last major action between 29 September and 1 October when the 5th and 3rd Australian Divisions and two American divisions attacked the Hindenburg Line across the top of the 6-kilometre-long St Quentin Canal tunnel; the canal was a major obstacle in the German defensive scheme.[9]

The citation for Campbell's Military Medal (MM) in the New Year's Honours List read.

> During operations from 29 September to 1 October 1918 in the Bellicourt Sector, Private Campbell displayed great gallantry and initiative. On 29 September he assisted to re-organise other troops who appeared to be leaderless, and then led his own section forward in the face of heavy machine gun fire. On the morning of 1 October immediately after the capture of Joncourt, he mounted the Church Tower and observed and reported the enemy movements under heavy shell and machine gun fire. During the three days he set a fine example of gallantry and devotion to duty.[10]

His two brothers Norman and Allen also served in WW1 and were each awarded a Military Cross for gallantry.

Following his service in the First World War, James became a plantation manager for Burns Philp and Co, in the Solomon Islands. Burns Philp and Co. was a major Australian shipping line and merchant that operated in the South Pacific area and had begun their operations on Bougainville in 1904. Campbell was awarded a soldier settlement grant of 2 hectares near Buka on Bougainville, where he and his wife Amy established a copra plantation. James aged 42, had married Amy Warburton a 25-year-old English nurse in August 1918, just before leaving to serve with the 32nd Battalion in France. A colleague of Campbells on Buka remembered him well and his excitement at the impending arrival of a special person. Of note in the article is that they had already been married in England but seem to have gone through the ceremony again in Bougainville.

> I used to look forward to Jim Campbell's visits as he was a good talker and used to tell me of his experiences in the earlier days. One day he arrived at my place from Buka Passage and informed me he was in Kieta [a port town on the east coast of

Bougainville] to await the arrival of his wife-to-be. She was a nurse whom he had met in England during the war. The bride duly arrived, and they were married by District Officer McAdam. The happy couple left Kieta immediately after the wedding ceremony to return to Jim's plantation. Their transport for the ninety-mile trip was an open cutter about twenty foot long. Mrs Campbell was a lovely woman and had a very hard life on the plantation. When Jim died, she carried on the plantation, which was still not bearing well, on her own. The chaps round Buka Passage were very good to her and helped her quite a lot.[11]

James had first been admitted to hospital with malaria in February 1918. He must have never quite recovered, as he passed away from blackwater fever or tertiary malaria at Kieta Hospital on 1 September 1930, aged 53 years.

Harry James CATTANACH (1886–89?)

Following his year of service in South Africa, he returned to Australia on March 1902, being awarded the QSA and entitled to five clasps for his service in the Cape Colony, Orange Free State, Transvaal and for South Africa 1901 and South Africa 1902. He was a member of the Committee which was formed to erect a memorial in Melbourne to those men of the VMR contingent who had died in the war. At the outbreak of the First World War, as a 34-year-old farmer [in fact he was closer to forty] from Jeparit, Harry enlisted on 31 August 1914 as a private with D Company of the 8th Battalion. He embarked for the Middle East, where he was part of the second wave to storm ashore at Anzac Cove on 25 April. On his first and only day in action, he was severely wounded on the beach and received a gunshot wound to his right thigh, a shrapnel wound to his left thigh and the loss of teeth because of a shrapnel blast. He was hospitalised in four different hospitals in England for five months until October 1915 and having lost the use of three toes on his right foot, returned to Australia in December 1915. His medical report from authorities in Melbourne recommended that he was unfit for further military service and that although his injuries were not permanent, the continual numbness and pain in his left leg caused difficulty in walking. Harry would always require ongoing massage and exercise to the muscles of his legs. The Army acknowledged that Harry had a total incapacity to earn a living and on 13 September 1916, he was discharged from the army as medically unfit and granted a war pension. One of nine children, he went to live with his mother and sisters in East Melbourne. Harry remained unmarried and died aged 50 at Ultima near Swan Hill on 23 November 1925.[12]

James William CHRISTIE (1885–86)
Died of enteric fever on 7 December 1900 at Rustenburg, South Africa.

Ernest Norman COFFEY (CGS Cadet Drill Instructor)
Repatriated to Australia for ill-health reasons and died of tuberculosis at Box Hill
on 18 September 1902 and was buried at the Box Hill Cemetery.

Arnold Mercer DAVIES (Malvern 1890s)
Arnold Mercer Davies was born in 1876 and aged 9 and 10 attended Melbourne
Grammar School in 1885–86. Malvern Grammar Schol was not opened until
1890 and he may have been a foundation student there as he won a speech night
prize in 1893, aged 17. Arnold Davies has a unique record among Australian
servicemen, having enlisted for military service in three wars. Following his service
as a private in the Boer War with A Squadron of the 3rd Victorian Bushmen's
Contingent from March 1900 until January 1901, he returned to Australia.
He lived in Melbourne and worked as a mine manager or metallurgical assayer.
The Davies' family were members of Melbourne's rich social class, and he was
married to Millicent Yuille in 1907 in a glittering social affair with a prodigious
list of expensive wedding presents. He and his new wife took up residence in East
Prahran. Arnold Davies aged 38, later enlisted for service in the First World War
and became a Sergeant in the 3rd Battalion of the Australian Naval and Military
Expeditionary Forces (ANMEF) 1914–1918. After the capture of Rabaul, this
Battalion of about 600 men was called 'Tropical Force' as it had been specially
enlisted for service in the tropics. Davies served for the remainder of the war in
New Guinea at places such as Morobe, Namatanai (New Ireland) and Kieta on
Bougainville. It was during this time that he collected and acquired numerous
tribal artefacts from New Guinea, which in turn he sold his collection of 23 items
to the South Australian Museum in early 1921. He returned to New Guinea,
probably prospecting for gold and his previous employment as a mineral's assayer,
no doubt led him to see some untapped possibilities in the areas he had visited
during his war service. In 1934 he was based in the Wewak area of the north coast
of New Guinea. At this time Wewak was the main supply port for the rich alluvial
gold fields at Maprik, some 70 kilometres to the west in the Torricelli Mountains.
Incredibly at the age of sixty-six, although he claimed he was only fifty-five years
old, Davies, named as a prospector, enlisted for army service ion World War 2 on
8 September 1942 at Randwick in Sydney. His enlistment papers indicate that
he was single and seem to indicate that he was estranged from his wife and two
children in Melbourne. The papers also showed that he could not drive a motor

ANMEF

Germany had set up a network of powerful telegraph and wireless stations across the south-west Pacific. They had been installed after Germany had 'taken possession' of numerous Pacific islands: among them were Nauru (1906), the German Solomon Islands Protectorate (1885), German Samoa (1900), German New Guinea (1884) and the Bismarck Archipelago (1885). These wireless and telegraph stations aided communications between the German East Asia Squadron's naval ships and land forces. When war was declared on 4 August 1914, Australia was asked to destroy the radio stations and to occupy German New Guinea and the surrounding areas. Recruiting for such a force began in earnest and by 20 August the Australian Naval and Military Expeditionary Force (**ANMEF**) force of 1,000 infantry from NSW and 500 naval reserves from Qld, NSW, Victoria and SA had been recruited, well-equipped and safely embarked. On the morning of Friday 11 September 1914, 25 naval reservists landed on the Bismarck Archipelago and after a small battle destroyed the radio station at Bita Paka. Six Australians were killed and four wounded in the action with Seaman W.G.V. Williams of Northcote becoming the first Australian fatality of the war. On Monday 13 September members of the ANMEF raised the British flag in Rabaul, the provincial capital. The forces of ANMEF, later 'Tropical Force' then began the Australian military occupation of German New Guinea which lasted until the end of the war. **Arnold Mercer Davies** served as a sergeant with this force for the war's duration.

car or ride a motorcycle but could cook and use a typewriter. However, some 26 days later, he was discharged as medically unfit for service. What happened to Davies after this is uncertain, but in August 1951, aged 75, while living in Sydney, he made a claim for a replacement for his ANMEF badge, which he claimed he had lost while crossing a river in New Guinea sometime in the period 1921–22.

Henry Gascoigne DAVIES (Malvern 1896? Scotch College 1897)

Following his service as a Trooper with the 2nd Scottish Horse in the Boer War, Henry worked as a law clerk. On the outbreak of the First World War, and in March 1915 he enlisted as a Private with the 24th Battalion and served at Gallipoli from August 1915 until the evacuation in December 1915. On 4 April 1916, he joined the recently formed 59th Battalion AIF reaching France on

29 June 1916, but aged 37 was killed in action at Fromelles in France on 19th
July 1916. The 59th Battalion lost 20 officers and 675 men during the fruitless
attacks on this day and the body of Davies was seen in 'No-Man's Land' but was
never recovered. His name is recorded on a memorial at V.C. Corner Australian
Cemetery Memorial, Fromelles, France.

Edward Arthur DUNCAN (1893–94?)

Edward (Ted) Duncan later married a sister of one of his Boer War comrades,
Agnes Mary Wallace in 1909. In early 1912 he became the owner of the Lilydale
Hotel in Victoria, joined the Lilydale Progress Association, became a vice-
president of the Lilydale Cricket Club, the President of the Lilydale Football
Club and showed an interest in politics. In March 1914, the local newspaper
proudly announced that:

> … to keep abreast of the times, Mr E.A. Duncan, proprietor of the Lilydale Hotel,
> has purchased a motor car, which, as advertised in this issue with a competent
> chauffeur, is for hire at all hours, especially to meet trains anywhere.[13]

He enlisted in the AIF in June 1915 and the hotel licence was transferred to his
wife Agnes Duncan. Promoted to Sergeant and aged 40, he was attached to the
6th Australian Machine Gun Company, but he was first wounded by shellfire
and then later killed in action near Ypres in Belgium on 26th September 1917.
A newspaper article of eulogy from his birthplace in Victoria's Alpine area talked
more about his life.

> Sergeant E.A Duncan (Ted) of Lilydale and formerly of Wandiligong has been
> killed in action on 26 September. He was well known and greatly respected, not
> only locally, but throughout the North-East. For many years he engaged in mining
> locally and at the time of the Klondike Rush [in northwestern Canada in the late
> 1890s] he tried that field. At the outbreak of the Boer War, Sergeant Duncan was
> amongst the volunteers who left this district, and, in that campaign, he gained
> distinction for dashing work. When the appeal was made for men in this State [for
> the Great War] Sergeant Duncan was conducting an hotel at Lilydale. He felt the
> call applied to him and after making business arrangements, he offered his services
> and soon gained promotion. Up to the time of his demise, although he had been
> in several hot engagements, he had always come through unscathed. 'Ted' who
> was known to everyone locally, was a man of genial disposition, a great favourite
> with a wide circle of friends and a gentleman in every sense of the word. At the
> last general election for the Legislative Assembly, 'Ted' unsuccessfully contested

the Evelyn seat in the Labor interest, [coming second in the result, but increasing Labor's vote by 10%]. His sympathies were always with the mining industry and although removed from Wandiligong, he was ever ready to champion his native township as a field for investment capital. He was one of the promoters of the Minerva [Gold] Dredging Company and was a large shareholder. In addition to his wife, Sergeant Duncan leaves a family of three sons and one girl for whom, as also for his mother and widow, deep sympathy has been expressed. As a mark of respect, the flag at the Shire Hall was flown half-mast on Thursday.[14]

Thomas Barham FOSTER (1890–91)

Died of enteric fever on 22 August 1901 at Umtali, Rhodesia (Zimbabwe).

Charles Reginald HANDFIELD (Malvern 1895–96?)

Handfield was made a Corporal in the Imperial Light Horse and in 1902 received the Queen's South Africa Medal and was also entitled to wear clasps for active service in the Orange Free State and the Transvaal. He returned to Australia in 1906 and again in 1912. His sporting prowess came to the fore, when in the 1908–09 season he played first class cricket for Transvaal in the Currie Cup, South Africa's equivalent to Australia's Sheffield Shield. He appeared in just one first class game when Transvaal played Border on 23 March 1909 and took two catches, but only scored 5 runs.

By the outbreak of the First World War in August 1914 Handfield was back in South Africa where he joined the Natal Light Horse (NLH), once again with the rank of Corporal. This unit of six hundred men was recruited in just ten days and comprised only men who had seen previous active service, as well as quite a few Australians. In what was seen as the first major Allied success of the entire war, the NLH was engaged in chasing the German Army from South African territory back to German controlled South-West Africa, now Namibia. Handfield was part of a 'flying column' which attacked the German rear guard at a small railway point called Gibeon Station on 27th April 1915, but in the short and fiercely fought action, nearly 30 NLH soldiers were killed. Amongst the casualties was Charles Handfield who died of his wounds on 6th May 1915 and was buried amongst his former comrades in the war cemetery at Gibeon Station. After the action the German armed forces retreated to South-West Africa never to return. Charles Handfield was the second Malvern Grammarian to be killed in the First World War.

Elmslie Fayrer HEWITT (1888 –?)

It is believed that Hewitt resumed farming at North Motton, south-east of Devonport in Tasmania, before moving to Queensland and enlisting in the 1st Division of the AIF in the First World War. His subsequent fate during this conflict was explained in a Hobart newspaper article in 1929.

DEATH PRESUMED. STRANGE CASE OF A SOLDIER WHO DISAPPEARED AFTER GALLIPOLI.

There were some interesting particulars in a matter mentioned by Mr. A. N. Lewis (Lewis, Hudspoth, Perkins, and Dear) before the Chief Justice (Sir Herbert Nicholls) In the Practice Court at Hobart yesterday. The matter concerned an application regarding letters of administration in the estate of Elmslie Fayrer Hewitt, and the motion was for the right to presume his death. In an affidavit Mr. Lewis set out that Hewitt, so far as could be discovered, left Australia with the 1st Division, A.I.F., from Queensland, in August or September 1914. He did not communicate with anybody prior to leaving the Commonwealth, and all that his people ever received was his photograph in uniform. It was believed that he was killed at the landing at Gallipoli on April 25, 1915. He was there, but there had never been any report of him being wounded or killed. He was not reported missing; there was just no trace of him at all. He just disappeared. It was a most remarkable thing that the case should be of a man who was at the Gallipoli landing and disappeared into thin air, without even being reported as 'missing.' Nobody could be traced who saw him on the day of the landing, and he had never been heard of since. It was quite easy, of course, that he should have been killed without any trace on such a day as that of the landing, but the peculiar feature of the case was that there was no report that he was missing.' Leave to presume the death', as asked, was granted.[15]

In 1917 on his older brother's WW1 military service file, Elmslie was listed as the Next of Kin, and was reported to be on 'Active Service – Whereabouts Unknown,' but there is no trace or record of his own Great War service file at the National Archives in Canberra. Only his Boer War service record is held on official records.

Reginald James John HOLLOWAY (1895–96)

Following his Boer War service and permanently affected in his right leg by the serious head wound he had sustained, Holloway later graduated in metallurgy at the University of Melbourne. He became an assayer with various mining companies, later setting up his own business as an assaying chemist in Collins

Hussey Burgh George Macartney pictured aged 40 as a Captain in the 2nd Royal Fusiliers in the First World War. (https://www.ukphotoarchive.org.uk/the-sphere-portraits-m)

THIS CROSS
FROM THE GRAVE OF
CAPTAIN HUSSEY BURGH GEORGE MACARTNEY.
ROYAL FUSILIERS.
AT LA BRIQUE. ST JEAN-LES-YPRES.
IS PLACED HERE BY HIS SISTER
MRS ROBERT HAYES. OF RADIPOLE MANOR
IN PROUD AND LOVING MEMORY.
"ONE OF MANY WHO PERISHED. NOT IN VAIN
AS A TYPE OF OUR CHIVALRY.

After the Great War, the original cross from George Macartney's battlefield grave at Ypres, was placed in St Anne's Church, Radipole, England, by his sister Emily. A brass memorial plaque was situated nearby, which to acknowledge his death read in part, 'One of many who perished, not in vain, as a type of our chivalry.'[18] (https://thereturned.co.uk/crosses/radipole-church-of-st-anne-dorset/)

Street, Melbourne. He soon returned to farming, firstly in Western Australia, then back to the family property at Tyntynder, near Boort. Reginald Holloway passed away in 1965 and is buried at the Swan Hill cemetery.[16]

Hussey Burgh George MACARTNEY (1886–92?)

After recovering from a serious head wound which passed from ear to ear, and incurred early in the Boer War, George Macartney was promoted to Lieutenant in 1899 and then Captain in 1904, before retiring from the Army in 1912. He took up farming on Vancouver Island in Canada between 1912 and 1914 but was visiting England at the time of the outbreak of the First World War in 1914. Answering the call of Empire, he enlisted in the 1st Regiment of his old battalion, the Royal Fusiliers, but aged 40, Captain Hussey Burgh George Macartney was killed in action near Ypres in Belgium on 24 June 1915.[17]

A W McLEAN (Malvern 1890s)

An A W McLean is recorded as being invalided home from Durban on the *Windsor* on 14th April 1901 and was visiting relatives in Australia.

He became an employee of the British Embassy in Japan working as a Shipping Clerk in Shimonseki in 1905–06, a Clerk and Accountant in Tokyo from 1907 onwards until 1921–23 when was registered as the Archivist and Accountant. In 1912 he visited Sydney, and his comments about Japanese maritime expansion were reported in the press.

> The Japanese Government is opening a steamship service from Japan to Singapore and then to open direct trade with Pacific Islands. The service would be at first confined to the Caroline and Gilbert Groups and their outlying islands. He does not think that Australia need fear them, but they [Japanese] are looking towards the South Seas Islands, and he does not see how they can be kept from going there.[19]

He did provide an article to the MGS school magazine, *The Malvernian* in 1907 in which he sets out his record in the Boer War and reminds his readers that as he was only nineteen at the time, he was unable to enlist in any Victorian contingent and subsequently travelled to South Africa to join up.

Of particular interest is that he is recognised as the individual who introduced Australian Rules Football to Japan in 1910 when he founded the Seisoka Football Club. In a letter to a Perth newspaper in 1910, enclosing a photograph of the Seisoka Football Club, he outlined his claim.

> I introduced the game to Japan this winter, and so far, the results have been better

than anticipated. Already four middle schools in this city are playing it and it is probable that more may be induced to take it up in the near future. I have been personally coaching them all and it has been encouraging to note the keenness they display. When I introduced the Australian game, I had the rules translated into Japanese and provided the various clubs with the necessary materials.[20]

In October 1923, McLean was one of the few Australians left in Yokohama after the devastating earthquakes in Japan on 1 September. It was reported at the time that the greater part of Tokyo lay in ashes with at least 142,800 people dead and 40,000 missing with about 30,000 houses destroyed in Tokyo alone. An Australian newspaper reported.

> Last night two members of the press delegation were stranded on shore due to the typhoon. It was dangerous to be out after dark owing to the sentries, but they were in company with Mr McLean, one of the few Australians left here, who still has brothers living in Melbourne. They walked through miles of ruin and mud, the scene of desolation, to the hills and safety.[21]

A W McLean is lost to the public record after this date.

Ronald Valentine Swanwater MACPHERSON (1896–98?)

When the First World War broke out, he enlisted as a Lieutenant in the 29th Field Artillery Brigade and saw service in France, was promoted to Captain and transferred to the 7th Field Artillery Brigade. Promoted to Major on August 1917, he was Mentioned in Despatches on 7th November 1917, 'for gallant service and devotion to duty in the field during most of 1917.' McPherson was severely wounded in action on 20th of November 1917, when he received gunshot wounds to his legs as well as suffering a fractured skull. He was discharged medically unfit from the AIF in March 1918. After a shipboard romance he married Ethel Strickland at All Saints Church, St Kilda on 30 September 1918. A newspaper article in October 1918 related the circumstances of Ronald McPherson's marriage as well proving some background to the career of his new wife.

> TROOPSHIP ROMANCE. ARMY MATRON
> AND AUSTRALIAN OFFICER.
>
> On Monday last Miss Ethel Maude Strickland, M.C., one of the military nursing sisterhood, changed her army uniform for conventional wedding attire to marry Major Ronald Valentine Swanwater McPherson, at All Saints' Church, St. Kilda. The bride was attended by Miss Jean Harmsworth and Miss McPherson; sister

of the bridegroom, Mr. Fred McCullagh (West Australia) gave her away. Later
Mrs. Reginald Harmsworth, sister of the bride, entertained many relatives of
both families at Wickliffe House. In honour of the bridegroom, the wedding cake
was decorated with his artillery colours, also miniature shells, and wagons. The
nurse and soldier first became acquainted on a vessel by which Major McPherson
returned to Melbourne a few months ago. Miss Strickland has been doing duty as
a matron on hospital ships since they were first commissioned. She enlisted as an
ordinary army nurse and was not aware that she had been appointed matron of the
outgoing boat until she went to headquarters to sign on. More than once, she had
been in close quarters with submarines, but risks such as these have not destroyed
her nerve. She confesses that her most trying experience was being placed under
the fire of the cinema operators in Hyde Park, when the King decorated her with
a Military Cross on June 2nd, 1917. Miss Strickland is a trainee of the Melbourne
Hospital, where she was a staff-sister for some time. She is a granddaughter of the
late Rev. F. Strickland, of Port Melbourne, who many years ago was connected
with the Aborigines Mission Station at Coranderrk. Major McPherson is a
relieving manager with the Colonial Bank.[22]

He then resumed his position in the Colonial Bank of Australasia, an event noted
in a 1938 newspaper article, under the heading 'The Three Members of the State
Economic Committee'. Late in 1938 Professor D.B. Copland agreed to the request
of the Victorian Premier Albert Dunstan, to chair a State Economic Committee
(S.E.C). Composed of A.T. Smithers, R.V.S. McPherson, and Copland, the
Committee had a threefold brief of reporting to the Government on economic
conditions in Australia and in Victoria and on the relationship between public
finance and employment. Dunstan referred to them other specific matters and in
1939 had the Committee undertake a major enquiry into the railways.

Mr RVS McPherson is manager of the Loan and Mercantile Co. Ltd, Melbourne.
He was born at St Arnaud and educated at Caulfield Grammar School. He began
his business career in the Colonial Bank of Australasia Ltd., and shortly afterwards
saw active service in the South African war with the Victorian Mounted Rifles.
On return to Australia, he resumed his career in the bank, being attached to
its staff in many branches. On the outbreak of the Great War, he joined the
Australian Field Artillery and was wounded in France. After the war, when the
Colonial Bank was amalgamated with the National Bank of Australasia, he
was appointed sub-manager of the head office, which position he held until he

took over the management of the N.Z. Loan and Mercantile Agency Co. Ltd., Melbourne.[23]

Ronald McPherson died aged 73 at Brighton on 12 August 1954.

Stanley Spencer REID (1886)

Died of wounds at Middelburg, South Africa on 23 June 1901.

George Frederick ROBERTS (1898 –?)

Roberts had enlisted in the 2nd Scottish Horse as Trooper no. 40137 on 2nd January 1902 and saw service in South Africa until 6th July 1902, before returning to Australia and therefore was entitled to the Transvaal clasp to his 1902 service medal. Little else is known of him after his war service.

Andrew Percival ROWAN (1889–90?)

Having seen service in South Africa for his mandatory period of enlistment, Percy Rowan returned to Victoria on board the *Custodian* on 26th April 1902. Following the actions and engagements of the 5th VMR, Percy was entitled to the QSA and five clasps for service in Cape Colony, Orange Free State, Transvaal, South Africa 1901 and South Africa 1902. He joined the militia and served with the Australian Field Artillery in Victoria from 1902–1906, being promoted to the rank of Lieutenant in July 1903 and was placed on the reserve list in 1910.

He took up farming on the 100,000-acre property at near Talleyrand near Longreach in Queensland but later relocated to West Australia with his brother to pursue grazing and farming

Lt Andrew Percival Rowan (https://www.caulfieldgrammarians.com.au/alumni-profiles/andrew-percival-rowan/)

CGS, MGS AND GALLIPOLI

More than one hundred Caulfield and Malvern Grammarians were involved in the Anzac Campaign between 25 April and 20 December 1915. Four Caulfield Grammarians, **Claude T Crowl, Leslie J Langdon, John Melvin** and **Henry E Walker** were killed on the first day as was **Erle F.D. Fethers** (Malvern). Caulfield Grammarian Boer War veteran **Harry Cattanach** was severely wounded. In May 1915 the British Army attacked the village of Krithia using Australian battalions as reinforcements. Among the Caulfield dead were **Herbert H Hunter** and **Robert W Langlands**. As in the Boer War, issues of poor sanitation and disease led to the deaths of **Vivian C Tricks** and **Frederick K Looker** (Malvern). **Gordon C Mathison,** a brilliant doctor was killed by a stray bullet behind the lines. **Norman Siddall** was killed in action as an artilleryman. **Cedric Permezel** (Malvern) died of wounds on a hospital ship. **Sherbourne H Sheppard** was killed by shellfire in trench operations. The 1904 Dux of CGS, **George Croker** was awarded a Military Cross for gallantry in action. The attacks on Lone Pine and The Nek on 7 August 1915 cost the lives of six Grammarians namely, **Kenneth R Brown, Colin Cramond, George Ormerod, Gordon McRae, Percy Rowan** and **William Groom. Horace C Harton** was killed in the torpedoing of the troop ship HMT Southland. **Bertram Atkinson** (Malvern) was killed in action at Lone Pine a month later. Of the 8,500 Australians who had died at Gallipoli, **21** were Caulfield or Malvern Grammarians.

activities in the area around Nungarin, some 280km east of Perth. Percy was recognised as having a splendid physique, was uniformly courteous and possessed a broad-minded manner. He took a keen interest in his local farming community and was a member of the Farmers and Settlers Association.

When the First World War broke out, he enlisted on 29 October 1914, as he was a Lieutenant in the Reserve and took this rank into the AIF. He was taken on strength with the 10th Light Horse, which left for Egypt on board HMAT Surada A52 from Fremantle on 17 February 1915. They landed at Alexandria and on 16 May 1915, embarked for Gallipoli. Although they belonged to a Light Horse Regiment, they fought as infantry in the Gallipoli campaign. Here, Rowan was in action for three months, suffering a slight wound to the neck in May, which did not need hospitalisation. On his recovery, Percy was appointed to the rank of Temporary Captain on 1st August 1915. Five days later the British planned to land further north at Suvla Bay and the AIF provided supporting attacks elsewhere on Gallipoli. On this afternoon the Australians attacked and occupied the Turkish front lines at Lone Pine against determined Turkish counterattacks. Caulfield Grammarian Private Kenneth Roy Brown (11th Battalion) was killed on 6th August. To further support the fighting at Lone Pine, an attack was planned along a narrow ridge known as The Nek. On the morning of 7th August four waves of Australians were cut down before they reached the enemy line. The Australian official historian referring to these Light Horseman, (LH) later wrote:

> A failure to synchronise watches meant that the Allied bombardment stopped seven minutes before the advance at 4.30am. As a result, the first line of 150 men from the 8th LH was swept away by Turkish fire almost as it rose up. No one in the line had not been knocked to the ground before advancing a mere 10 yards in front of the trench. The second line, also from the 8th LH and the third line from the 10th LH were fully aware that they were almost certainly going to their deaths, but they did not hesitate. Their valour was not enough, and they were also cut down in the same manner as the first line. In an hour 650 men and officers out of a force of 800 were killed or wounded. The flower of the youth of Victoria and Western Australia fell in that attempt.

The Official History describes the circumstances of Rowan's death at The Nek. Rowan was in the Fourth line and a halt had been called to the futile charges, whilst further orders were sought but:

> ...the impression was somehow created that the charge had been ordered. The troops on the right at once leapt out and instantly there burst forth the same (Turkish) tempest of machine gun fire. The nearest NCOs looked at Captain Rowan their troop leader, who signed to them to go, at the same time rising himself and waving his hand, only to fall back dead from the parapet. His troop sergeant Sanderson repeated the signal and the men in the centre sprang out.[24]

Another member of the regiment recalled the circumstances.

> We had been given the order to leave the trench and charge. Captain Rowan was a big, tall man, and the trench was very deep, so I had cut him two steps in the side, and we gave him a leg up. He was the first over the parapet and called out, 'Come on boys.' He was hit immediately on the forehead and in the lung by a machine gun. His own batman [Corporal Moore] and I held him up after he had been hit. We pulled him down into the trench and held him up and a doctor was brought at once, for he was a good officer, and we thought the world of him. The doctor

did all he could for him, but it was hopeless, and he died very soon afterwards. He was a splendid chap, and was always, the first out of the trench.[25]

Amongst the dead at The Nek were five Caulfield Grammarians. Trooper Colin Hearder Cramond 8th Light Horse (LH), Trooper George Booth Ormerod (8th LH), Trooper Gordon McRae (10th LH), Captain William Edward Groom (14th Battalion) and Captain Andrew Percival Rowan (10th LH). These futile charges at The Nek were later immortalized in the 1981 film *Gallipoli*.

A Melbourne newspaper report said of Rowan:

> The deceased officer, who was 39 years of age, was well and favourably known in Melbourne, being of magnificent physique, and of sterling qualities.[26]

Finally, a newspaper from Queensland where he had previously farmed noted that.

> Lieutenant Rowan, who was killed in action recently at the Dardanelles, was a brother of Mr J B Rowan of Talleyrand and Tandilla. The late Lt Rowan was only 39 years of age, and he was well known out here having resided on Talleyrand on several occasions since the conclusion of the Boer War, through which he went without receiving a scratch. He received his Lieutenant's commission in South Africa and went to the Dardanelles from West Australia and was twice wounded and the third time he was shot dead.[27]

His widowed mother, Violet Elinor Bowen, was listed as his next of kin, living at Cliveden Mansions, Wellington Parade, East Melbourne, in 1915.[28]

Walter Laishley SPIER (1888)
Died of enteric fever on 23 January 1901 at Woodstock Hospital, Capetown, South Africa.

George Raleigh STEWART (1886 -?)
George returned to Australia, was discharged on 10th October 1902 and was entitled to the QSA and three clasps for service in the Transvaal and South Africa (1901) and South Africa (1902). Little else is known of him after his Boer War service.

Thomas STOCK (1893–96)
Killed in action on 12 February 1900 at Hobkirk's Farm, Pink Hill, Rensburg, South Africa.

Charles Hugh THOMAS (1890)

After his Boer War service Charles qualified for the QSA and five clasps for his work in Cape Colony, Orange Free State, Transvaal, South Africa 1901 and South Africa 1902. He returned to Melbourne on board the *Custodian* on 26th April 1902. He became an overseer and took up work with the Pacific Phosphate Company (PPC) on Ocean Island and whilst there he served for five months as a Sergeant Major and 2nd Lieutenant in the Ocean Island Volunteer Reserve.[29] With the outbreak of the First World War and upon his return to St Kilda, he enlisted as a gunner in December 1915 with the Field Artillery Brigade (FAB) and was later appointed as a Lieutenant in January 1916. Posted to the 9th FAB in April 1916, he served in France and was recommended for a Military Cross, but instead was Mentioned in Despatches for gallantry in 1918. After the war he returned to live in St Kilda and married Gwendoline Elizabeth Story in 1928 and then went back to work on Ocean Island. As World War 2 spread, in February 1942 a French destroyer evacuated the PPV employees, including Charles Thomas, as Japanese forces occupied the island from 26 August 1942 until the end of the war. Charles returned to Australia but died at Richmond on 19th February 1949.

Frank Valentine WEIR (1893)

Upon the declaration of the First World War in 1914, Frank Weir enlisted in the Australian Light Horse and saw action at Gallipoli, where he was wounded and with the Egyptian Expeditionary Force where he was again wounded twice. He married Dorothy Isherwood at St Anne's On Sea, England on 2nd October 1916 and then as a Major was posted to the First Light Horse Brigade HQ. In August 1917 he was admitted to the 14th General Hospital in Egypt suffering from bronchitis, but by the following month was headed back to England as he was 'flourishing.' He saw action on the Western Front and

Frank Weir pictured in uniform during the First World War. (https://www.caulfieldgrammarians.com.au/alumni-profiles/frank-weir/)

was awarded the Distinguished Service Order (DSO) on 25th July 1918, for acts of gallantry and devotion to duty in the field and was also Mentioned in Despatches (MiD).

After the war he resumed his farming activities in the Deniliquin area becoming a well-known grazier at Bertangles, Bowning. In addition, he became a Trustee of Church Property for the Anglican Diocese of Riverina, and during World War 2 he held the rank of Colonel in the Volunteer Defence Corps at Yass. He died in February 1966 and is buried at Deniliquin, NSW.

Gerald Massey Ivor WILKINSON (1895–96)
Killed in action on 3 July 1901 at Elandskloof, South Africa.

Arthur George Thomas WILLIAMS (1888–?)
Little is known about Arthur after his Boer War service, although he is recorded as having married Agnes Maud Black in 1918.

Gerald Massey Ivor Wilkinson

Endnotes

1   Droogleever. R. *That Ragged Mob.* Trojan Press. Thomastown, Victoria. 2009. pp529–30.

2   http://broomeanzacs.broomemuseum.org.au

3   *Bendigo Advertiser.* Wednesday 10 September 1902. p5.

4   The Indian Ecclesiastical Establishment had been founded in 1813 under the auspices of the East India Company to train and provide Anglican priests for India.

5   *Mercury* (Hobart). Friday 20 February. 1914. p4.

6   Droogleever. R. *A Matter of Honour.* 2017. Trojan Press, Thomastown, Victoria. p306

7   Copra is the dried, white flesh of the coconut from which coconut oil is extracted.

8   The Buka Passage is a narrow strait that separates Buka Island from Bougainville Island.

9   https://www.awm.gov.au/collection/U51472

10 *Commonwealth Gazette.* No. 109 15 September 1919.

11 *Una Voce* (Journal of the PNG Association of Australia Inc.) No.1 March 2004. p15.

12 https://emhs.org.au/biography/cattanach/harry

13 *The Lilydale Express and Yarra Glen, Wandin Yallock, Upper Yarra, Healesville and Ringwood Chronicle.* Friday 13 March 1914. p4

14 *Alpine Observer and North-Eastern Herald.* Friday 12 October 1917. p2

15 *Mercury* (Hobart, Tasmania) Tuesday 5 March 1929, p5.

16 Droogleever. R. *That Ragged Mob.* ibid. p397

17 *Alumni Cantabrigienses: A Biographical List of All Known Students, Graduates and Holders of Office at the University of Cambridge, from the Earliest Times to 1900. Volume 2. From 1752–1900. Part 4. Kahlenburg – Oyler.* Cambridge University Press. 1951. edited by John Venn. p251.

18 The line of poetry is taken from the poem called 'By Flood and Field,' by the Australian poet Adam Lindsay Gordon.

19 *Huon Times* (Tasmania). Saturday 3 February 1912. p6.

20 *Mirror* (Perth). Friday 20 May 1910. p9.

21 *Advertiser* (Adelaide). Monday 15 October 1923. p12.

22 *Shepparton Advertiser* (Vic), Monday 7 October 1918. p1.

23 *Age* Wednesday 31 August 1938, p13.

24 Bean. C.E.W. *The Official History of Australia in the War of 1914–18.* Volume II. p619.

25 'An Officer's Death'. *Argus* (Melbourne). Tuesday February 1916. p5.

26 *Argus* Wednesday 25 August 1915. p7

27 *Northern Miner* (Charters Towers) 13 September 1915. p7

28 https://emhs.org.au/biography/rowan/andrewpercival

29 Ocean Island is now called Banaba Island and is the western most point of Kiribati in the Pacific Ocean. It has an area of just 6 square kilometres and with its highest point being 81 metres was one of the most important elevated phosphate rich islands of the Pacific. It has no natural streams, and the traditional source of water was a cave near the sea. From 1900 and for the next 80 years, the British Phosphate Commission, equally owned by Australia, New Zealand and the UK, mined Banaba so extensively that about 90% of the island's surface was stripped bare. (https://www.theguardian.com/world/2021/jun/09/the-island-with-no-water-how-foreign-mining-destroyed-banaba

CHAPTER 10

Epilogue

It could be asked, why would Caulfield Grammarians living in another country on another continent, be involved in a military conflict so far from home, with potential for long-lasting implications. For some it cost them their lives; others were to live out the remainder of their lives with the consequences of debilitating wounds.

The participation of Australians in the Boer War, and later in the First World War, was seen in some quarters as an exercise in the defence of the British Empire. It could not be claimed in October 1899 that Australia was under military threat of invasion from the Boer Republics, but Australians certainly viewed themselves as part of the British Empire and were prepared to share the load if menace threatened. At the end of the nineteenth century and the outbreak of the Boer War, writers extolled the virtues of the British Empire and saw that its citizens were all members of one imperial body; a body that would unite in a common defence of its laws, liberties and values.[1] Australians saw themselves as independent 'Australian Britons' and defenders of the Empire in the Pacific, who if called upon, would volunteer to give their lives in defence of the ideals of that Empire. It was said, that in time through hard work and the application of imperial traditions, Australia [and other English-speaking colonies] would become better than the old British nation in Europe and be a part of the empire on which the sun never sets.[2]

School atlases, books and newspapers showed the advance of the Empire and there was a distinct national pride in Australia in the Empire's achievement and vision, with essentially no conflict seen between the interests of Great Britain and Australia.

While it was seen that Australia's place in the Empire was important in a variety of ways, it had no great military history of its own. Small numbers of Australian volunteers had served in the Māori Wars earlier in the century, and some troops from New South Wales had taken part in the Sudan Campaign of the early 1880s. No large-scale Australian contingents or expeditionary forces had ever been sent overseas before to serve the Empire. Our young country's soldiers had not really been tested in battle against a foreign foe. However, the doctrine of military unity of the Empire was widely accepted and potential young volunteers were being prepared by society and at schools in several ways.

Nineteenth-century Australian schools had been largely modelled on the great schools of England, which had become the benchmarks for educational and social success across the world. In addition to a focus on the world of academics and classics, three key factors underpinned the patriotic fervour that ensured high rates of enlistments across all colonies once hostilities were declared in 1899.

The first factor was an underlying and 'all-pervading' imperial ethos to all parts of the curriculum. Indeed, just as British school boys had been trained to do their duty by the Empire, Australians and our Grammarians followed suit. No Australian flag existed until 1901, no unique Australian National Anthem existed until 1984 and with atlases and maps coloured in imperial red, society and schools were heavily laden with an imperial ethos. Britain, its flag, its anthem, its reverence for royalty and all things 'from home in England' dominated all Australian public occasions, thinking and school life. The 'all-pervading role' of the British Empire in the lives of all Australians, was writ large for example, in the Oath of Attestation signed by each Australian soldier upon enlistment for the Boer War. This piece of paper made it clear exactly where the priorities and duties of the colonial soldier lay. It was signed by the soldier, and then by a witness in front of a Justice of the Peace who finally added his own signature. It read as follows:

> I *[full name]* do make oath that I will be faithful and bear true allegiance to Her Majesty, Her heirs and successors, and that I will, as in duty bound, honestly and faithfully defend Her Majesty and successors in person, Crown and dignity against all enemies, and will observe and obey all orders of Her Majesty, her heirs and successors, and of the Generals and Officers set over me. So, help me God.

It can be noted that not one mention of Australia was made anywhere in this document.

Secondly, as outlined earlier, students were strongly encouraged to be involved in sport and athletic pursuits. This rise in 'athleticism' was justified, because it was argued that organised games not only helped produce fit young men but also generated school spirit and taught boys fair play, teamwork and how to win and lose graciously. Sport was seen as a preparation, a training ground for some higher ideal and that he who succeeded in sport had equipped himself well to lead in politics, business, the professions or perhaps even the military. Boys were encouraged to see these activities as a chance to learn about the great game of life and how to rule the state and in turn the Empire. Schoolmasters began to believe that in time, if the great game was to be war, it would become an easy transition from loyalty to one's house or school to clearly direct that devotion towards a soldier's regiment. Boys were taught that success in war depended upon patriotism and military spirit.

In Victoria, headmasters, such as L. A. Adamson of Wesley College emphasized the importance of excellence in sport and extolled the effects on character of training and preparation for military service in his students and claimed:

> The British love of games had proved a magnificent asset to the Empire producing unselfish, devoted leaders, able to endure hardship and discomfort. Schoolboys might well continue with their games, even in times of war, because sport equipped them to take their place at the front if this was necessary.[3]

Thirdly, the introduction of school cadet units in the 1860s in some Australian boys' schools had initially been facilitated by the provision of large playing fields at boys' schools which, apart from being used for sporting pursuits, were also used on a regular basis for military drill for cadets. With the withdrawal of the permanent British Army regiments from Australia in 1870, the newly established Australian colonial military forces saw school cadets as a valuable recruiting ground. Headmasters also viewed cadets as a vehicle to further educational and disciplinary goals, while some sections of the community saw the organisation as helping to further positively develop the character of young boys and youth. The CGS Cadet unit was established in 1885 and all the first three CGS Headmasters strongly supported its role and activities. Lt. Col. Frederick Sargood was for a time the Victorian Minister of Defence and owned 'Ripponlea', the large suburban property in the vicinity of CGS, with two of his sons being Grammarians. Sargood had been instrumental in establishing school cadets in Victoria and at the founding meeting in March 1884 stated that he saw cadets as furnishing a most important recruitment ground for the militia forces. His overriding goal in founding school cadets was made quite clear when he stated that he intended:

> To bind together in one patriotic brotherhood the youth of this country so that, should occasion arise, they may be able in after years to defend their country with the most telling effects.[4]

In 1902 Sargood gave a newspaper interview about cadets and said how pleased he was to note that many students had passed through its ranks and that it appeared that more than two-thirds of the men who had served in the five Victorian Contingents in South Africa had been school cadets. He believed this accounted for the readiness with which they picked up the drill and rifle work once they entered camps for their training.

In addition to the influences from their schooling, there is no doubt that the young Grammarians who volunteered for military service, as with other volunteers in wartime, did so because of the promise of travel, the undertaking of a heroic adventure, mateship, sheer excitement and the fact that they were determined not to be kept out of it. Still others sought a regular source of income and a chance to travel and, in this instance, to see South Africa. Like all soldiers, they would have known that death or severe wounding might await them, but with the optimism of youth, this was not a deterrent. This is illustrated by the fact that the early death of Tom Stock, just two months after his arrival at the war in 1900, did not seem to deter future enlistments. In the case of several former students, they enlisted together in the same regiment, while a number also served a second 'tour' of duty with different regiments.

While Australians saw themselves as the inheritors of proud British military traditions, there was still some deeply held concern about whether our soldiers would prove worthy of that tradition. Australia, in essence was not just an emerging young Federation in the early 1900s, it felt itself under scrutiny, as its fighting men had yet to be really tested in a major conflict. In some measure these Grammarians and their fellow soldiers, unwittingly or otherwise, set out to make a name for their country and themselves, by serving their Empire well in South Africa, given the limited impact that 16,500 Australian troops could make in the context of a British Army of some 450,000 men. It was well recognised that the Australian mounted men were the best such troops that the British Army had in the Boer War. However, there was still a prevailing feeling that Australians had not really been put to the test in battle and questions remained about their steadfastness and' battle-hardness.' This question was to be answered in full measure on the shores of Gallipoli thirteen years later.

The twenty-four Caulfield Grammarians and four Malvern Grammarians who enlisted for military service in the Second Anglo-Boer War had grown up in a country and society where love of Empire was seen as a paramount virtue, because it linked the young, soon to be federated nation of Australia to a vast world-wide Empire. The headmasters who oversaw the instilling of key values through their school's curriculum of academics, sport and cadets, passionately believed in their critical roles in shaping young lives. CGS Headmasters Davies, Barnett and Buntine championed this approach in their leadership of the growing educational institution in East St Kilda and ensured that their young students' spiritual, academic and physical needs were well met.

In conclusion, with the passage of time from those Boer War days, we can see the Grammarian volunteers in military service reflected the context of school and society in which they grew to manhood. As with their country, they were young, enthusiastic, easy-going Australians who wanted to serve their school, Queen, country and regiment, and to fight and acquit themselves bravely and to never let down their comrades. The record shows that they were prepared to only rest contentedly when they had consistently laboured hard and well.

Endnotes

1 Robertson J. *ANZAC and Empire – The Tragedy and Glory of Gallipoli*. Hamlyn Australia. 1990. Melbourne. p10.

2 Robertson J. ibid. p11.

3 McKernan M. *The Australian People and the Great War*. Nelson Australia. 1980. Melbourne. p101

4 Stockings C.A.J. *The Torch and the Sword*. The History of the Army Cadet Movement in Australia. UNSW Press. Sydney, 2007. p33.

Appendix

NOMINAL ROLL OF CGS AND MGS BOER WAR SOLDIERS

NAME		CGS Years	No.	Rank	Unit	Died	Notes	Other school
Anderson	Andrew Percival	1884–89?	333	Pte	4th Victorian Imperial Bushmen		Survived and travelled to England before RTA	
Bardwell	Bernard Everett	1892–96?	-	Lt	6th West Australian Mounted Infantry		Survived enteric fever and RTA	
Birch	Ormonde Winstanley	1896–98?	1491	Sgt	5th Victorian Mounted Rifles		Survived and RTA	
Boyd	William	1886–90?	479	Pte	4th Victorian Imperial Bushmen		Survived and RTA	
Campbell	James William	1884–91?	-	Pte	6th Imperial Bushmen (NSW)			
			2275	L/Cpl	4th Australian Commonwealth Horse		Survived and RTA	
Cattanach	Harry James Goodrich	1886–89?	1686	Pte	5th Victorian Mounted Rifles		Survived and RTA	
Christie	James William	1885–86	491	Pte	3rd Victorian Bushmen	7/12/1900	Died of enteric fever at Rustenburg	Melbourne Grammar School
Coffey	Ernest Norman	Cadet Staff	1	CSM	1st Victorian Mounted Infantry	18/9/1902	RTA. Died of tuberculosis at home in Box Hill.	
Davies	Arnold Mercer	MGS -?	454	Pte	3rd Victorian Contingent		Survived and RTA	
Davies	Henry Gascoigne	MGS -?	31579	Trpr	2nd Scottish Horse		Survived and RTA	

NAME		CGS Years	No.	Rank	Unit	Died	Notes	Other school
Duncan	Edward Arthur	1893–94?	493	L/Cpl	4th Victorian Imperial Bushmen		Enlisted under name of a brother, James Thomas Duncan. RTA	
Foster	Thomas Barham	1890–91	367	Pte	4th Victorian Imperial Bushmen	22/8/1900	Died of enteric fever at Umtali	Ballarat College
Handfield	Charles Reginald	MGS 1895–6	1578	Cpl	1st Imperial Light Horse (South African)		Survived and RTA	
Hewitt	Elmslie Fayrer	1888 -?	170	Pte	2nd Tasmanian Imperial Bushmen		Survived and RTA	
Holloway	Reginald James	1895–96	268	Pte	4th Victorian Imperial Bushmen		Survived dangerous head wound. RTA	
Macartney	Hussey Burgh George	1896–922?	-	2nd Lt	2nd Battalion Royal Fusiliers (British Army)		Survived dangerous head wound. RTUK	
Mclean	A.W.	MGS -?	-	Cpl	South African Light Horse		Survived and RTA	
Mcpherson	Ronald Valentine Swanwater	1896–98?	1304	Cpl	5th Victorian Mounted Rifles		Survived and RTA	
Reid	Stanley Spencer	1886 (Feb to April)	41	Pte	2nd WA Mounted Infantry		Wounded in stomach. RTA	Scotch College
			-	Lt	6th WA Mounted Infantry	23/6/1901	Died of wounds at Middleburg	
Roberts	George Frederick	1898 -?	40137	Trpr	2nd Scottish Horse		Survived and RTA	
Rowan	Andrew Percival	1889–90?	76	Pte	5th Victorian Mounted Rifles		Survived and RTA	

NAME		CGS Years	No.	Rank	Unit	Died	Notes	Other school
Spier	Walter Laishley	1888	152	Cpl	NSW Citizen's Bushmen	23/1/1901	Died of Enteric Fever at Woodstock Hospital, Capetown	Sydney Grammar School
Stewart	George Raleigh	1886 -?	36892	Trpr	1st Scottish Horse		Survived and RTA	
Stock	Thomas	1893–96	89	Pte	1st Victorian Mounted Rifles	12/2/1900	KIA at Hobkirk's Farm, Pink Hill, Rensburg	
Thomas	Charles Hugh	1890	840	Pte	5th Victorian Mounted Rifles		Survived and RTA	
Weir	Frank Valentine	1893	15	Sgt	NSW Citizen's Bushmen		Survived and RTA	
			–	Lt	1st Australian Commonwealth Horse			
Wilkinson	Gerald Massey Ivor	1895–96	31757	Trpr	2nd Scottish Horse	3/7/1901	KIA at Elandskloof	
Williams	Arthur George Thomas	1888 -?	213	Pte	4th Victorian Imperial Bushmen		Survived and RTA	

KEY

Pte – Private
CSM – Company Sergeant Major
RTA – Returned to Australia
KIA – Killed in Action

L/Cpl – Lance Corporal
Lt – Lieutenant
MiD – Mentioned in Despatches
MGS – Malvern Grammar School

Cpl – Corporal
2nd Lt – 2nd Lieutenant
RTUK – Returned to United Kingdom

Sgt – Sergeant
Trpr – Trooper

BIBLIOGRAPHY

BOOKS

Alumni Cantabrigienses: A Biographical List of All Known Students, Graduates and Holders of Office at the University of Cambridge, from the Earliest Times to 1900. Volume 2. From 1752–1900. Part 4. Kahlenburg – Oyler.

Australian War Memorial. *South African War 1899–1902.* Campaign Series No. 2. Progress Press Canberra. 1976.

Barthrop, M. *The Anglo-Boer Wars. The British and the Afrikaners 1815–1902.* Cassell. London 1991.

Bean, C.E.W. *The Official History of Australia in the War of 1914–18.* Volume II. 1940.

Brown, D. *Athleticism in Australia: St Peter's College, Adelaide: A Case study in the diffusion of a Victorian Educational Ideology.* unpublished paper, University of Queensland, Dec. 1986, p7. cited in Mangan. *ibid.*

Campbell, J. A. (Lt. Col.) (ed) *History of Western Australian Contingents serving in South Africa during the Boer War (1899–1902)* (HWAC) Government Printer, Perth, 1910.

Clark, M. *A History of Australia. The People Make Laws.* Volume V. MUP. Carlton. 1987.

Clements, M.A. *Adamson, Lawrence Arthur (1860–1932)* Australian Dictionary of Biography, volume 7. (MUP). 1979.

Coulthard-Clark, C. *The Encyclopedia of Australia's Battles.* Allen and Unwin. Crow's Nest. NSW. 2001.

Doyle, A.C. *The Great Boer War.* Nelson and Sons. London. 1903.

Grant, R.G. (ed). *1001 Battles that changed the course of History.* Harper Collins Books. Sydney. 2011.

Droogleever, R. *Colonel Tom's Boys. Being the Regimental History of the 1st and 2nd Victorian Contingents in the Boer War.* PrintBooks. South Melbourne. 2013.

Droogleever, R. *That Ragged Mob – Being the Service Record of the 3rd and 4th Victorian Bushmen Contingents in the Boer War.* Trojan Press. Melbourne. 2009.

Droogleever, R. *A Matter of Honour. Being the History of the 5th Contingent of the Victorian Mounted Rifles in the Boer War, 1901–1902.* Trojan Press. Melbourne. 2017.

Haley, B. *The Healthy Body and Victorian Culture.* Cambridge: Harvard University Press. 1978.

Hanson, N. *The Unknown Soldier. The Story of the Missing of the Great War.* Doubleday. London. 2007.

Hobshaw, E.J. *The Age of Empire 1875–1914.* Weidenfeld and Nicolson Ltd. 1987.

Horner, D. *Strategy and Command – Issues in Australia's Twentieth-Century Wars.* Cambridge University Press. 2021.

Hutchison, G. *Remember Them. A Guide to Victoria's Wartime Heritage.* Hardie Grant Books. Prahran. 2009.

Inglis, K.S. *Sacred Places – War Memorials in the Australian Landscape.* MUP. 2008.

James, L. *The Rise and Fall of the British Empire.* London. 1982.

Kenyon, J.P. *The Wordsworth Dictionary of British History.* Market House Books Ltd. 1981.

King, J. *Great Battles in Australian History.* Allen and Unwin. Sydney. 2011.

Lloyd, B. and Nunn, K. *Bright Gold – The story of the People and the Gold of Bright and Wandiligong.* Histec Publications. 1987.

McIntosh, P. *Physical Education in England since 1800.* (London, 1968), p70. in Mangan *ibid.*

McKernan, M. *The Australian People and the Great War.* Nelson Australia. Melbourne. 1980.

Mangan, J.A. *Manufactured Masculinity. Making Imperial Manliness, Morality and Militarism.* (Sport in the Global Society – Historical Perspectives) Routledge. Abingdon. 2012.

Mann, J. and Allen. D. *Fallen. The Ultimate Heroes. Footballers who never returned from War.* Crown Content. Melbourne. 2002.

Meyer, A.R. CSKLS Annual General Meeting. Baylor University. 2010. Journal of Sport and Social Issues.

Murray, P.L. *Official Records of the Australian Military Contingents to the War in South Africa.* 1911.

Pakenham, T. *The Boer War.* Folio Society edition. London. 1999.

Penrose, H. *Outside the Square. 125 Years of Caulfield Grammar School.* MUP. 2006.

Price, J.E. *Southern Cross Scots.* BR Printing. Kensington. Vic. 1992.

Robertson, J. *ANZAC and Empire – The Tragedy and Glory of Gallipoli.* Hamlyn Australia.1990. Melbourne.

Sherington, G. *Athleticism in the Antipodes: The AAGPS of New South Wales.* History of Education Review, 12(2). 1983.

Smurthwaite, D. *The Boer War 1899–1902.* Hamlyn. London. 1999.

Steiner, Z and Neilson K. *Britain and the Origins of the First World War.* (2nd Ed). Palgrave MacMillan. Hampshire. UK. 1977.

Stockings, C.A.J. *The Torch and the Sword. The History of the Army Cadet Movement in Australia.* UNSW Press. Sydney. 2007.

Una Voce (Journal of the PNG Association of Australia Inc.) No.1. March 2004.

Webber, H. *Years May Pass On. Caulfield Grammar School 1881–1981.* Wilke and Company Clayton. 1981.

Wilkinson, I. *The Fields at Play – 115 years of Sport at Caulfield Grammar School 1881–1996.* Playright Publishing. Sydney. 1997.

Wilmot, P. (Ed) Tom Stock Letters published in the *'Casterton News.'* Quoted from *'The Second Harvest – The Writings of Tom and Duncan Stock, two Victorian brothers in the Boer War, 1899–1902.'* 2000.

NEWSPAPERS

The Advertiser (Adelaide)

AGE (Melbourne)

Alpine Observer and North-Eastern Herald

ARGUS (Melbourne)

Australasian

Bendigo Advertiser

Brighton Southern Cross

Casterton News (Vic)

Feilding Times (NZ)

The Herald (Melbourne)

Huon Times (Tas)

The Lilydale Express and Yarra Glen, Wandin Yallock, Upper Yarra, Healesville and Ringwood Chronicle.

The Maryborough Advertiser

The Mercury (Hobart)

Maryborough Argus

Mirror (Perth)

Mornington Standard

New Zealand Times

Northern Miner (Charters Towers)

Ovens and Murray Advertiser (Vic)

Richmond Guardian (Melbourne)

The Shepparton Advertiser

The Sydney Mail and New South Wales Advertiser

The Farmer and Settler (NSW)

The West Australian

Weekly Times

Western Mail (WA)

SCHOOL SOURCES

Caulfield Grammar School magazine. *The Cricket.* October 1888. Vol.1.

Caulfield Grammar School magazine. (various and as referenced)

Caulfield Grammar School Jubilee Book. 1881–1931.

Caulfield Grammar School Speech Night programs (various).

Davies, J.H. *Personal Diary*. Caulfield Grammar School Archives.

Durie, M. *Caulfield Grammar School's Founder's Day Address*. 2008.

Melbourne Grammar School magazine. *The Melburnian*.

Ross, P. M. *Letter to D.J. Moran*. 13 December 1992. (Held in the Caulfield Grammar School Archives)

Sydney Grammar School magazine. *Sydneians in the Boer War*. SGS Winter 2016.

The Walter Murray Buntine Acquisition. File 8 – Item 0039: Caulfield Grammar School Prospectus 1883.

The Walter Murray Buntine Acquisition. File 2 – Item 017: Letter from Rev. E. J. Barnett to CGS Parents. 23 March 1896. Caulfield Grammar School Archives.

DIGITAL SOURCES

https://alh-research.tripod.com/Light Horse/Index.blog/topic id-1115696 3rd Victorian Bushmen. Outline.

https://www.angloboerwar.com/unit information/imperial units

https://angloboerwar.com/unit-information/south-african-units/345-imperial-light-horse

https://www.awm.gov.au/collection/U51472

https://www.awm.gov.au/collection/U52011

https://www.awm.gov.au/collection/U52012

http://www.bwm.org.au/contingents.php

https://www.bwm.org.au/soldiers/Hobart_Cato.php

https://emhs.org.au/biography/cattanach/harry

https://emhs.org.au/biography/rowan/andrewpercival

www.familyhistory.co.uk/the-boer-war pp5-6

nla.gov.au/nla.news-article10555156

nla.gov.au/nla.news-article33212201

http://alh-research.tripod.com/LightHorse/index.blog?topic_id=1113235 Victorian (Volunteer) Mounted Rifles

https://samilhistory.com/2023/10/26/the-boer-war-myth-busting-by-the-numbers/

http://samilitaryhistory.org/vol096jh.html

OTHER SOURCES

Commonwealth Gazette. No. 109. 15 September 1919. p1368.

Droogleever, R. Email to Dr Daryl Moran in October 2019.

Victorian Government Gazette. No. 25. March 6. 1885. p710.

Weir, F.V. *Personal Diary* held in Mitchell Library, Sydney.

INDEX

ABOUT THE AUTHOR

Dr Daryl Moran attended Caulfield Grammar School (CGS) from 1966–1970 and was a member of the School Committee, 1st XVIII and held senior ranks in the CGS Cadet Unit. Pursuing a career in education, he was appointed as a foundation staff member to Caulfield's Wheelers Hill campus in 1981, later becoming the Head of the Junior School. In 1990, he took up the position on the School's Executive as the Director of Development, being responsible for all matters relating to 'Friend-Raising' and 'Fundraising', including being Executive Director of both the CGS Foundation and the school's alumni body, the Caulfield Grammarians' Association (CGA). His brother taught at CGS for over 25 years, both his children attended the Wheelers Hill Campus, his son-in-law is a current staff member, and his five grandchildren are also Caulfield Grammarians. For several years, his wife, Jenny was the President of the CGS Society of the Arts. In 2024, Daryl was awarded the Don Wirth Medal by the CGA in recognition of his service to the school and the Association over many years.

He held Headships at other Melbourne schools before taking up a two-year teaching position as a foundation staff member at an international school in India, and holds a PhD from Melbourne University.

The son of a wartime member of the RAAF, Daryl has had a long interest in military and aviation war history, convening conferences concerned with Australian air power for Military History & Heritage Victoria Inc. He has been a Guest Speaker for MHHV and organisations such as the RSL and various historical societies. In addition to being a Guest Speaker at numerous CGS Anzac Day assemblies, he delivered the keynote address for Boer War Day at Melbourne's Shrine of Remembrance in 2024.

His first military title, *Empire's Noble Son* (2019), is the biography of 2nd Lt 'Lyle' Buntine, MC, a Caulfield Grammarian who flew in the Royal Flying Corps in World War I. In collaboration with Dr Andrew Kilsby, Daryl co-edited and contributed chapters to the book *In the Fight* (2024), which outlines the story of Australian involvement in the Burma Campaign during World War II. Their next book, *Resolute – The Australian Air War in the Burma Campaign 1942–45* (2025), tells the story of 1700 RAAF aircrew who served in RAF Squadrons in the Burma Theatre. They are also collaborating with an international writing team on the book *Allied with India – Australians in the Indian Army 1800–1947* to be released in 2026.

Daryl is a long-standing member of Rotary International, being a former Club President, current District Chair of Youth Service, and a Past District Governor.

www.ingramcontent.com/pod-product-compliance
Lightning Source LLC
Chambersburg PA
CBHW041012140426
R18136400001B/R181364PG42813CBX00009B/7